W9-BCK-613

CONTRARIES

CONTRARIES

ESSAYS

Joyce Carol Oates

New York
OXFORD UNIVERSITY PRESS
1981

Library of Congress Cataloging in Publication Data

Oates, Joyce Carol, 1938-
Contraries.

1. English literature—History and criticism—
Addresses, essays, lectures. I. Title.
PR403.017 820′.9 80-21429
ISBN 0-19-502884-8

Printed in the United States of America

for my colleagues in criticism
at Princeton, Rutgers,
New York University—

Preface

Without Contraries is no progression.
William Blake

Literary criticism, the most ingenious form storytelling can take.

What is it, for all its rituals of language, its sacred codes, its systematic reflections upon reflections, but the protracted effort to define and accommodate the powerful emotions stirred by works of art? Criticism speaks, as Northrop Frye has observed, and all the arts are silent. But the reverberations of that silence—!

The critic is a pilgrim, an acolyte, a translator; a gnostic intermediary fueled by the need to bring metaphors from one system to another. He analyzes, traces, compares, elucidates, composes "arguments" . . . in the hope of dealing, in intellectually negotiable terms, with the troubled nature of our relationship to art.

That it is a troubled relationship seems to me incontestable. We are stimulated to emotional response not by works that confirm our sense of the world, but by works that challenge it. (To "identify with"—a commonplace pietism—means simply "to have no further thoughts about.") The "Contraries" of which Blake speaks in *The Marriage of Heaven and Hell* are in fact the very energies of Blakean "Delight"—the Eternal Delight that rests in motion, in strife, in passion, in revolutionary violence. Energy, as Blake says, is the *only life*.

The seven essays in this volume, written over a period of approximately twenty years, and intermittently revised, were originally stimulated by feelings of opposition and, in two or three cases, a deep and passionate revulsion. That one is stirred, excited, baffled, and even upset by a work of the magnitude of *King Lear* goes without saying; that one might also be angered by it, annoyed, vexed, offended, and even depressed—and even, during a solid month's immersion in its ambagious poetry, made insomniac—is perhaps less readily admitted. "Classics," after all, are works most people no longer question in primitive emotional terms.

The English and Scottish "traditional" ballads, which I first studied in my early twenties, struck me initially as too simple for literary criticism—for the modernist New Criticism fashionable at that time, with its elegant apparatus for "dissecting" opacity in poetry suitably opaque. *Ulysses*, which I first attempted at the age of sixteen, struck me, for all its wonders of language (which leap off the page, even to the uninstructed eye) as too willfully complex, too self-indulgent, a monument of egotism—my mistake, but not mine alone. D. H. Lawrence was shrill and hectoring and "obvious"; and embarrassing too, in his insistence upon the primacy of the "blood"—for what, after all, *is* the "blood"? Oscar Wilde offended with his cute paradoxes, his cloying epigrams, his worked-over prose; Dostoyevsky, though clearly a master, was so flagrantly and smugly and (it seemed) spitefully Christian . . . a bias that deformed the great novels. These initial responses, this sort of immediate opposition, had at least the advantage for me of provoking me to thought: for a quarrel with others, in Yeats's famous but still useful definition, may lead to rhetoric: but "rhetoric" itself may lead to something more valuable.

Why must Cordelia die, one asks, *why must Conrad murder Martin Decoud—the man so much like himself!* It has never been a fashionable critical technique, to my knowledge, to fantasize structures for works of art other than those they contain; but the practicing writer knows how eerily fluid these structures are, at the very outset of a work—how open to improvisation, audacity, hope—before the artist's secret life-drama, the private mythology working itself out through his imagination, intervenes in the guise of aesthetic necessity. So the question *Why* is asked, *Why this and not that*, impulsive questions, bold, quixotic, *Why this fate and not that?* —fueled by the spirit of contrariety that lies at the heart of all passionate commitment.

Princeton, N.J.
October 1980

J.C.O.

Acknowledgments

" 'In the Fifth Act': The Art of the English and Scottish Traditional Ballads" consists of two essays, one of which appeared in *Dalhousie Review*, Winter 1963, and the other in *Southern Review*, Summer 1979.

"Is This the Promised End?" appeared in *The Journal of Aesthetics and Art Criticism*, Fall 1974.

"Tragic Rites in Dostoyevsky's *The Possessed*" appeared in *The Georgia Review*, Fall 1978.

" 'The Immense Indifference of Things': Conrad's *Nostromo*" appeared in *Novel*, Fall 1975.

"*The Picture of Dorian Gray:* Wilde's Parable of the Fall" appeared in *Critical Inquiry*, Summer 1980.

"Lawrence's Götterdämmerung: The Apocalyptic Vision of *Women in Love*" appeared in *Critical Inquiry*, Spring 1978.

"Jocoserious Joyce" appeared in *Critical Inquiry*, Summer 1976.

Contents

CONTRARIES

The Picture of Dorian Gray: *Wilde's Parable of the Fall*

Its parable-like simplicity and the rather painful remorselessness of its concluding chapters have made it possible for readers to underestimate the subtlety of *The Picture of Dorian Gray*. So clearly does its famous plot move to its ineludible climax—so explicitly are its major "points" articulated (the poisonously charming Lord Henry is told: "You cut life to pieces with your epigrams")—that the complexity of Oscar Wilde's imagination is likely to be minimized. While in one sense *The Picture of Dorian Gray* is as transparent as a medieval allegory, and its structure as workman-like as that of Marlowe's *Dr. Faustus*, to which it bears an obvious family resemblance, in another sense it remains a puzzle: knotted, convoluted, brilliantly enigmatic: and if one might be "poisoned by a book" (as poor Dorian charges he has been, by Huysmans's rather silly novel), *The Picture of Dorian Gray* might very well be that book.

Joyce saw in Wilde not only a fellow-artist but a betrayed artist and a "dishonored exile"—a kind of Christ—though his initial response to *The Picture of Dorian Gray* was qualified: the book was "crowded with lies and epigrams" and its spirit muted by the fact that Wilde felt obliged to "veil" the homosexual implications.[1] That Joyce was insensitive to Wilde's deeper theme is suggested by his frequent echoing of Wilde ("My art is not a mirror held up to nature," Joyce boasts. "Nature mirrors my art"), which develops only the explicit, daylight side of Wilde's aesthetics and makes no allusion to the cautionary and even elegiac tone of much of *The Picture of Dorian Gray*. It is not, certainly, the homosexual nature of Dorian's behavior—or, for that matter, his allegedly promiscuous heterosexual behavior—that constitutes Dorian's sin, but the fact that he involves others in his life's drama "simply as a method of procuring extraordinary sensations," and without emotion. Joyce's imagination was earthbound, even domestic, and if Stephan Dedalus with his ashplant and his pose of languid weariness brings to mind the adolescent defiance of Wilde's infamous "The Decay of Lying" of 1889, it is the case nonetheless that Joyce's artist was hardly likely to drift into satanism, let alone murder. Nighttown, in fact, will make a victim of *him*.

Beyond the defiance of the young iconoclast—Wilde himself, of course—and the rather perfunctory curve of *Dorian* to that gothic final sight (beautiful Dorian dead with a knife in his heart, "withered, wrinkled, and loathsome of visage") there is another, possibly less strident, but more central theme. That one is damned for selling one's soul to the devil (for whatever prize—"eternal youth" is a trivial enough one) is a commonplace in legend; what arrests our attention more, perhaps, is Wilde's claim or boast or worry or warning that one might indeed be poisoned by a book—and that the artist, even the presumably "good" Basil Hallward, is the diabolical agent. Wilde's novel must be seen as a highly serious meditation upon the moral role of the artist—an interior challenge, in fact, to the insouciance of the famous pronouncements that would assure us that there is no such thing as a moral or an immoral book ("Books are well written, or badly written. That is all"), or that all art is "quite useless." Wilde's genius was disfigured by his talent: he always sounds much more flippant, far more superficial, than he really is. So one is always saying about *Dorian,* with an air of surprise, that the novel is exceptionally good after all—and anyone who has read it recently replies, with the same air of faint incredulity, yes, it *is* exceptionally good—one of the strongest and most haunting of English novels, in fact. Yet its reputation remains questionable. Gerald Weales virtually dismisses it as "terribly *fin de siècle*" in his rather flippant introduction to the Signet paperback edition, and it would be difficult to find a critic who would choose to discuss it in terms other than the familiar ones of Decadence, Art-for-Art's-Sake, Art as "the telling of beautiful untrue things."

Beneath the entertaining and often distracting glitter of Wilde's verbal surfaces, however, one does discover another work; or if not precisely another *work,* then another tone, and another Wilde. Suddenly life is not a matter of dialogue, of drawing-room repartee; it takes on the nostalgia of "unutterable longing" which Pater found in Shakespeare (see Pater's *Appreciations* of 1889), and that ceremonial, elegiac rhythm one senses in the great plays: as if Shakespeare had had in mind some "inverted rite" by which human justice executes its sentences. Just as Pater translated Shakespeare's drama into other, more static forms—the tapestry, the religious rite, sculpture, pictorial art—so Wilde translates, or transmutes, what might be in another era a tragedy of the violent warring of consciousness with itself into a reassuringly old-fashioned morality play. The ceremonially correct punishment of Dorian Gray seems to complete the novel; but in fact it merely ends the novel. The preoccupation with the questionable morality of the artist's interference with life—Basil Hallward's appropriation of Dorian's image, for instance, for his uncanny portrait—is never satisfactorily resolved, and even the final appearance of the aging and somewhat attenuated Lord Henry hints at another level of human concern which Wilde has no space to investigate. What the strangely moved reader is likely to carry away from *Dorian* is precisely this sense of something riddling and incomplete. One feels about it as one feels about the most profoundly haunting works of art—that it has not been fully understood.

The murder of Basil Hallward by Dorian Gray is usually seen as one of the more demonic of Dorian's acts. Yet the murder is symbolically appropriate, and appropriate too is the fact that, for Dorian, this former idolator ("I worshipped you too much") becomes a loathsome "thing" after his death and must be eradicated by crude scientific means. (Cut up, presumably, and dissolved with nitric acid in a sleight-of-hand Wilde feels no need to make plausible.) Basil functions as a "good" character, one of Lord Henry's straight men, but his role in Dorian's damnation is hardly an ambiguous one, and his sudden death answers to an internal logic.

"Actual life was chaos," Dorian thinks, as his moral disintegration allows him insight, "but there was something terribly logical in the imagination": by which Wilde suggests the limits of Lord Henry's (and his own) faith in the power of the individual will. To become a spectator of one's own life, as Henry has boasted, not only fails to save one from suffering—it makes suffering inevitable, although the "suffering" of course will come in unanticipated forms, as ennui, paralysis, the "shallow moods and sickly thoughts" of protracted adolescence. Dorian's wickedness appears to be involuntary; he would not have exchanged his soul for eternal youth and beauty had not an artist, Basil Hallward, presented him with an image of himself utterly new, unrequested, and irresistible—if, that is, the terrible logic of the imagination had not set into play a tragic sequence of events of which "Dorian Gray" happens to be the central figure.

There is no doubt but that Basil Hallward initiates the tragedy, for it is his worshipping of the young man's physical beauty and his appropriation of his image (as "art") that calls Dorian's attention to himself—and stimulates Lord Henry's undisguised homoerotic interest. (To Henry, Dorian is a young Adonis who looks as if he were made of "ivory and rose-leaves." He is a Narcissus, a brainless beautiful creature "who should always be here in winter when we have no flowers to look at.") Basil, however, is deeply troubled by his painting. He understands instinctively—despite his friend's gibes—that he must not exhibit it in public because it reveals too much of himself. *Why* he has painted Dorian Gray's picture is not clear, given his ambivalent feelings about Dorian's beauty and the fatality he believes attends all human beings of distinction. He says, in words that must echo Wilde's own thoughts on the subject, and on the very creation, in prose, of Dorian: "Every portrait that is painted with feeling is a portrait of the artist, not of the sitter. The sitter is merely the accident, the occasion." Though Basil cannot know that his dream-image will destroy him, quite literally, he has experienced, at their first meeting, one of those violent and inexplicable spasms of emotion that attend a "fatal" attraction. The role of the artist, so extravagantly proclaimed by Wilde as self-determined, and superior to mere life, sounds here as if it were a matter of involuntary fate. Basil felt terror in Dorian's presence: "I knew that I had come face to face with someone whose mere personality was so fascinating that, if I allowed it to do so, it would absorb my whole nature, my whole soul, my very art itself." But Basil's resistance is, of course, futile. He does give away his soul, he does fall in love with a boy who symbolizes the harmony of soul and body, and soon comes to feel that he could not live as an artist without Dorian.

Then again he says, quite bluntly, that Dorian is "simply a motive in art," and that his likeness on canvas might bear little resemblance to the young man himself. He reduces, in Basil's aesthetic imagination—overheated as it is—to a manner of painting, to "the curves of certain lines" and the loveliness and subtlety of "certain colors." Dorian has no soul or worth of his own; he functions as the artist's Muse or Anima, and his value lies in his unconscious (and feminine) stimulation of the male artist's energy. Basil Hallward is not in love with Dorian Gray but with his own image of Dorian, which is to say his own "motive" in art. Wilde knows that the artist oscillates between the frenzy of inspiration and the lucidity of an almost impersonal wisdom when he has Basil speak in this way, and to allude to a godly or supra-human destiny that will involve both the artist and his mesmerizing subject. Basil is fated to single out Dorian for his art and by means of his art to force Dorian into a tragic self-consciousness: by appropriating the boy's image in answer to an artistic motive he begins the boy's destruction, and it is altogether fitting that Dorian should murder him some years later. (As, it might be argued, Oscar Wilde's "artfulness" came close to destroying him, and did expel him forever from the society of presumably normal people.)

In a far less graceful fashion Hawthorne explored the problematic relationship between an artist of sorts and his subject, in the cautionary allegory "The Birthmark." Here, a scientist of genius attempts to remove the defect of a birthmark on his beautiful wife's cheek, with the inevitable result that the woman dies. In this ponderous tale there is no doubt that the scientist is the villain. He interferes with beauty, his "fatal hand" dares to grapple with "the mystery of life," and so an angelic spirit is loosed from its mortal frame. Hawthorne's Aylmer is an awkward Yankee cousin of Faust whose experience is more blundering than wicked; Hawthorne ends, somewhat obscurely, with the remark that had Aylmer reached a more profound wisdom "he need not thus have flung away the happiness which would have woven his mortal life of the selfsame texture with the celestial."

By contrast the fateful union of Basil and Dorian is predetermined by the gods. Basil, as the artist, *must* succumb to his motive; he *must* be seduced by the Adam-like Dorian, whose erotic powers are of course entirely unconscious. (Before meeting Lord Henry, and his own likeness, Dorian is indeed a brainless beautiful creature. He is somewhat spoiled, but spoiled in a child-like way; he is good-natured, spontaneous, and generous, an absolute innocent.) But the artist takes his image from him and exhibits it to him as art, as an object for contemplation. Dorian *objectifies* his own physical being, and his corruption begins at once: "When he saw [the portrait] he drew back, and his cheeks flushed for a moment with pleasure. A look of joy came into his eyes, as if he had recognized himself for the first time. . . . The sense of his own beauty came on him like a revelation." And immediately the boy begins to think, in language not unlike that of Lord Henry's, that his predicament is a very unhappy one. The surprise of the scene is that he *thinks* at all, still less in such a manner:

> "I shall grow old, and horrible, and dreadful. But this picture will remain always young. It will never be older than this particular day of June. . . . If it were only the other way! If it were I who was to be always young, and the picture that was to grow old! For that—for that—I would give everything! . . . I am jealous of everything whose beauty does not die. I am jealous of the portrait you have painted of me. Why should it keep what I must lose? . . . Why did you paint it? It will mock me some day—mock me horribly!"

Dorian's background, revealed later to Lord Henry, is suitably romantic: his mother was a beautiful young woman who ran away with a penniless subaltern, and who died shortly after his birth. The boy's misfortune charms Lord Henry, for beyond every exquisite thing there must be something tragic, "worlds had to be in travail, that the meanest flower might blow." The charm of Dorian is precisely his awakening from innocence to a realization of his own power. Talking to such a person, Henry thinks, is like playing upon an exquisite violin—and it strikes him as highly desirable that *he* should seek to dominate Dorian as Dorian, without knowing it, dominates Basil. There is something enthralling to Henry in the exercise of influence: "To project one's soul into some gracious form . . . ; to hear one's own intellectual views echoed back to one . . . ; to convey one's temperament into another. . . ." So Basil's "subject" becomes Henry's.

The Picture of Dorian Gray is a curious hybrid. Certainly it possesses a "supernatural" dimension, and its central image is gothic; yet in other respects it is Restoration comedy, energetically sustained for more than two hundred pages. It approximates the novel Lord Henry would write if he had the ambition—a book as lovely as a Persian carpet, and as unreal. The supernatural element, however, is never active except in terms of the portrait. It would be quite ludicrous if introduced to Lord Henry's drawing-room society, and it is really inexplicable given the secular nature of Dorian's personality. Evidently diabolical powers are stirred by Basil's art but Basil himself has no awareness of them, apart from a certain uneasiness regarding the morality of his relationship with Dorian. Is the Devil responsible? But does the Devil exist? Hell is hardly more than theoretical to Wilde, and Heaven is equally notional; when Dorian is attracted to the Catholic Church it is primarily for the sake of exotic ritual, ecclesiastical vestments, and other somewhat ludicrous treasures of the Church, which Wilde delights in cataloguing. The *consequences* of a Faustian pact with the Devil are dramatized, but the Devil himself is absent, which suggests that the novel is an elaborate fantasy locating the Fall within the human psyche alone. Basil, Lord Henry, and Dorian are all artists, aspects of their creator. Basil is a conventional artist, apart from his attraction to Dorian; Lord Henry and Dorian are aesthetes—artists of their own lives—whose hope is to generate a New Hedonism. Wilde echoes Pater, of course, but with a unique intensity, and it is significant that there is nothing remotely supernatural about such passages as the following, which are clearly at the very heart of the novel:

> There are few of us who have not sometimes wakened before dawn, either after one of those dreamless nights that make us almost enamoured of death, or one of those nights of horror and misshapen joy, when through the chambers of the brain sweep phantoms more terrible than reality itself, and instinct with that vivid life that lurks in all grotesques, and that lends to Gothic art its enduring vitality. . . . Veil after veil of thin dusky gauze is lifted, and by degrees the forms and colors of things are restored to them, and we watch the dawn remaking the world in its antique pattern. The wan mirrors get back their mimic life. . . . Nothing seems to us changed. Out of the unreal shadows of the night comes back the real life that we had known. We have to resume it where we had left off, and there steals over us a terrible sense of the necessity for the continuance of energy in the same wearisome round of stereotyped habits, or a wild longing, it may be, that our eyelids might open some morning upon a world that had been refashioned anew in the darkness . . . a world in which the past would have little or no place, or survive, at any rate, in no conscious form of obligation or regret. . . . It was the creation of such worlds as these that seemed to Dorian Gray to be the true object . . . of life.

Certainly it will not do to dismiss such sentiments as "romantic"—they strike a very deep chord, and underlie all creations of alternate worlds in art. If it is the case that the hedonist, unrestrained by morality, customs, or the surveillance of his neighbors, always drifts to the most extreme experiences, then it might be hypothesized that the hedonist is an archetype of man, not perverse but representational. His quest for new sensations in order to give vitality to his flickering life is nothing more than an exaggeration of any quest for meaning.

Wilde's contempt for the "shallow psychology" that defines man in terms of his social and familial position is as savage as D. H. Lawrence's. How is it possible, Dorian wonders, to conceive of man's ego as "a thing simple, permanent, reliable, and of one essence"? Wilde, like Lawrence, sees man as "a being with myriad lives and myriad sensations, a complex multiform creature that bore within itself strange legacies of thought and passion, and whose very flesh was tainted with the monstrous maladies of the dead." Insincerity, Wilde wittily observes, is merely a method by which we multiply our personalities. Where in Wilde the experimentation with alternative lives is a desperate means of escaping that *taedium vitae* that "comes upon those to whom life denies nothing," in Lawrence it is clearly quite different. Lawrence, who believed that "nothing that comes from the deep, passional soul is bad, or can be bad,"[2] was repulsed by the kind of hedonism Wilde preached, and he created the gnomish Loerke of *Women in Love* as a spokesman for *fin de siècle* aesthetics—with something of Italian Futurism thrown in. To the bat-like but highly articulate Loerke, art is purely self-referential: it is "a picture of nothing, of absolutely nothing," with no relationship to the everyday world, which exists on a separate and distinct plane of existence. Loerke is the "rock-bottom" of all life, the perfect stoic who is also the perfect epicure, troubled about nothing, connected to no one, a parody of Decadent aesthetics as it shades into European Fascism. That Loerke crouched in Lawrence's soul and fascinated him as he does Gudrun goes without saying. Even Lawrence's hatreds sprang out of sympathy.

Yet it would be wrong to assume that Lawrence's insights are inevitably deeper than Wilde's, or that his vicious portrait of Loerke ("I expect he is a Jew—or part Jew," Birkin mutters) is a more critical portrait than that of Dorian himself. Wilde's great theme is the Fall—the Fall of innocence and its consequences, the corruption of "natural" life by a sudden irrevocable consciousness (symbolized by Dorian's infatuation with himself)—but this falling from grace is available only to those who have attained a certain degree of economic and intellectual freedom. Restlessness, ennui, the inability to apply one's strength to anything—these are not merely symptoms of Dorian's perverse nature, but symptoms of a highly advanced and sophisticated civilization. So Dorian is a victim—not unlike Dostoyevsky's similarly emblematic Stavrogin, who drifts into a life of unimaginative vice because he is "freed" of the earth and of the necessity to labor as ordinary men do. Stavrogin is accursed by boredom: "As to my political views," he says in his confession, "I just felt I'd have liked to put gunpowder under the four corners of the world and blow the whole thing sky-high—if it had only been worth the trouble."[3] Stavrogin's role in the death of a young girl is more convincing than Dorian's role in the death of Sybil Vane, just as Stavrogin's predicament, generally, is more convincing in realistic terms than Dorian's: but both young men, handsome as they are, with the power to move others deeply while remaining unmoved themselves, are allegorical figures whose fates are meant to symbolize the sterility of an "advanced" civilization. How, Stavrogin asks, is one to apply one's strength, one's "limitless" strength, to anything that is not an illusion?

Dostoyevsky, no less than Wilde, is enamoured of his creation, but quite serious about the "strength" that might be applied to sheer labor, peasant labor. Stavrogin might have been redeemed had he knelt down to kiss the earth he had defiled, as Raskolnikov does; he might have saved himself had he followed Shatov's advice to "find God through labor." But Wilde would have none of this. He can be as sentimental as Dostoyevsky, and as unreasonably pious, but his "lower classes" are never anything but lower. They appear, in fact, to belong to a level of consciousness distinctly different from that of their elegant masters—tip-toeing about with their "masks of servility," lapsing into un-welcome "garrulousness" about tiresome household matters when their masters are eager to begin a night of vice. Two kinds of human beings, two species—those who are "free" and those who are not.

Dorian's freedom, however, as we know, is a consequence primarily of his loss of humanity. His soul is no longer his own: it has been appropriated by art. His response to Sybil Vane's melodramatic death is one of surprise and alarm at his own failure to feel grief: "Why cannot I feel this tragedy as much as I want to?" he complains to Lord Henry. "Real life" is eclipsed by art, and by the emotional responses we commonly give to works of art, rejoicing in their artificiality. The girl's death has for Dorian "all the terrible beauty of a Greek tragedy, a tragedy in which I took a great part, but by which I have not been wounded." In the end Henry explains Dorian's egotism to him, and assures him that Sybil Vane never lived—not for *him*—apart from the phantom in his imagination.

After the girl's death Dorian becomes increasingly detached from what might be called normal human emotions. It is interesting to note that the Shavian ideal—the man of disinterested sensibility who looks unmoved upon the melodrama of life—does not greatly differ from the precious, rather infantile, and supremely confident model of the dehumanized personality Wilde offers. Dorian is a golem, a parody of Lord Henry, assuring Basil Hallward that it is only shallow people who require years to get rid of an emotion: "A man who is master of himself can end a sorrow as easily as he can invent a pleasure. I don't want to be at the mercy of my emotions. I want to use them, to enjoy them, to dominate them."

Is it inevitable that a doctrine of Art-for-Art's Sake reduces to the sort of sickly, simpering aestheticism of a Des Esseintes? Is there something doomed in the very notion of a purely self-referential art that, at the very most, *uses* materials from the "real world"? One encounters repeatedly in literary history the belligerent claim that art has nothing to do with anything beyond itself, and that writing that aspires to the loveliness of, let us say, a Persian carpet must necessarily be "unreal." The stylist is encouraged to cultivate his own sensibility, for where actual life is surrendered to chaos one might nevertheless forge a certain logic of the imagination. In *Dorian,* Wilde surely believes in his aesthetics and at the same time offers, by way of Dorian and his fate, and Basil Hallward and *his* fate, a disturbingly prescient commentary on his beliefs: the artist who succumbs to the spell of Beauty will be destroyed, and so savagely that nothing of him will survive. The novel's power lies in the interstices of its parable—in those passages in which the author appears to be confessing doubts of both a personal and an impersonal nature.

It might be hypothesized that the airless and claustrophobic world of self-referential art, when it is not primarily a reaction against prevailing norms of "social realism" in its various guises, is actually a paradigm of the infant's world. Everything in that world is self-referential: everything refers inward: words, as they are grasped one by one, appear to be *created* by the child (just as the mother and other adults, mobilized by cries of hunger or alarm, certainly appear to be controlled by the child). The illusion of possessing and controlling everything is a powerful one, and its charms are not readily surrendered even in adulthood. So we encounter in theoreticians of self-referential art both the puzzling contempt for "real" worlds and the sentimental hope for a forcible remaking of the universe as if there were not already a universe to be acknowledged. The impulse for such creation—Faustian in its aspirations—must spring from a sense of insignificance; for even the infant's delusion of omnipotence is compensatory to its actual helplessess. *It is the spectator, and not life, that art really mirrors.*

The value of Wilde's allegory lies in the questions it asks rather than in the experience it transcribes. For *Dorian* gives us hardly any experience at all—it is surface and symbol, and too tidily constructed. Dorian, Lord Henry, and Basil Hallward fade, but their voices remain, asking certain unanswerable questions that are as appropriate for our time as they were for Wilde's: Is the Fall from innocence inevitable? Is the loss of illusion tragic, or comic, or merely farcical? Is the artist by his very nature inclined to manipulate and pervert his subject?—and is his doom bound up with the fact of his artistry, his autonomy? Lord Henry declares that if a man treats life artistically, his brain is his heart, but Wilde's novel—and Wilde's experience—suggests otherwise.

Notes

1. See Richard Ellmann, *James Joyce* (New York, 1959), p. 283 and p. 241.
2. See the Introduction to *Women in Love,* written in 1919.
3. Dostoyevsky, *The Possessed,* trans. Andrew R. MacAndrew (New York, Signet edition, 1962), p. 418.

Tragic Rites in Dostoyevsky's The Possessed

Somehow it has happened—no one knows quite how, or why—that the incidence of robbery and violence has doubled. Arsonists' fires have ravaged towns and villages, and in some places there is even disease: plague, and the threat of a cholera epidemic. The manager of a factory in the town of Shpigulin has shamelessly cheated the workers, and working conditions are very poor; subversive leaflets have appeared, urging the overthrow of the existing order; the idle, prankish company that routinely gathers in the Governor's mansion is becoming involved in adventures of an increasingly reckless kind. (They are called the Jeerers or the Tormentors.) The historic Church of the Nativity of Our Lady is plundered and a live mouse left behind the broken glass of the icon. Fedka, the escaped convict, a former serf who was sold into the army, many years before, in order to pay his master's gambling debt, roams the countryside committing crimes—not just robbery but arson and murder as well. The police seem unable to find him. "Strange characters" appear—a human flotsam that comes out of nowhere to plague society. Madmen erupt. Women become obsessed with feminism. Generals transform themselves into lawyers, divinity students speak out rudely, poets dress themselves in peasant costumes. The son of the province's most wealthy landowner has contracted a marriage—in jest, it would seem, after a night of drinking—with a woman of the very lowest social order, who is both lame and demented. A nineteen-year-old boy has committed suicide and a party of pleasure-seekers crowds into the room to examine him: one of the ladies says, "I'm so bored with everything that I can't afford to be too fussy about entertainment—anything will do as long as it's amusing." It seems that a number of people in the area have taken to hanging and shooting themselves. Is the ground suddenly starting to slip from beneath our feet? Is the great country of Russia as a whole approaching a crisis? Demons begin to appear, licking like flames about the foundations of order; a Trickster-Demon springs out of nowhere and, very much like the gloating Dionysus of Euripides' *The Bacchae,* wants only to sow disruption, madness, and death. "We shall proclaim destruction," Peter Verkhovensky tells his idol Stavrogin, "because—because . . . the idea is so attractive for some reason! And anyway, we need some exercise."

The Possessed, Dostoyevsky's most confused and violent novel, and his most satisfactorily "tragic" work, began to appear in serial form in 1871, close after the publication of *The Idiot,* and only a few years after the publication of *Crime and Punishment* in 1867. All of Dostoyevsky's great novels show a family resemblance, just as his marvelous operatic characters are obviously kin and might, without much difficulty, stride from one novel to another; but the demonic excesses of *The Possessed* seem to have sprung from the "plague" of which Raskolnikov dreams at the very conclusion of *Crime and Punishment,* when he is imprisoned in Siberia, a confessed but not truly repentant murderer. In a delirium Raskolnikov dreams that the world is condemned to a new plague from Asia, and that everyone is to be destroyed except a very few. The disease attacks men by way of their sanity: though mad, each believes that he alone has the truth and is estranged from his fellows. They cannot decide what is "evil," they do not know whom to blame, and they kill one another out of senseless spite, as the infection spreads. "Only a few men could be saved," Raskolnikov dreams. "They were a pure chosen people, destined to found a new race and a new life, to renew and purify the earth, but no one had seen these men, no one had heard their words and their voices."[1]

So in *The Possessed* madness is loosed. Society approaches a crisis: the classes freely intermingle in the Governor's mansion; the fires burn; ludicrous "revolutionary" theorizing gives way to action; the Fairy-Tale Prince, Stavrogin, commits suicide; and his spiritual father, his former tutor Stepan Trofimovich Verkhovensky, strikes out upon the road in a futile, desperate pilgrimage to "find Russia," and also dies. And there are other deaths, some of them bitter losses to Russia, indeed: Shatov and his wife Mary, and her newborn baby (who is Stavrogin's unacknowledged son); the inarticulate, mystical Kirilov; the wealthy young society woman Liza Drozdov; Maria Lebyatkin, Stavrogin's secret wife, and her brother Captain Lebyatkin; even the escaped convict Fedka. So wholesale a sweep of destruction suggests a tragic crisis, a violent erasing of old gods, old profanations, in order that the new may be born. Stavrogin's act of suicide is the act of magnanimity he had not believed within his grasp—he *does* become a hero, a Fairy-Tale Prince, not through the desultory acts of his largely wasted life but through the atonement of his death. By leaving the artificial paradise of his twenty years with Mrs. Stavrogin, Stepan Verkhovensky too becomes heroic: he is Don Quixote to Stavrogin's Hamlet. (Though Stavrogin is compared—wrongly, foolishly—to Prince Hal.) Both men have been superficially attracted by the "new ideas" that are plaguing

Russia, largely from the West; both are idle, even parasitic; they stand apart from the accelerated grotesqueries of the victimized community, but are unable, or unwilling, to prevent the impending catastrophes. Like everyone in *The Possessed,* with the possible exception of the near-anonymous narrator Govorov (who shares the curious unimaginative self-righteousness of the narrator of *The Brothers Karamazov),* Stavrogin and his former tutor are the most significant victims of a highly complicated saturnalia. Now mythic, now merely local (and cranky: for surely Govorov's acidulous comments on the "scum" are Dostoyevsky's own), legendary and historical by turn, *The Possessed* builds powerfully to one climax after another, ending with a festival of misrule that, in Durkheim's terms, will serve to revitalize the diseased cultural order. A stagnant social order drifts toward decadence, and decadence plunges into chaos: but chaos in itself can be a ritual experience, a pathway of redemption. For out of the primeval chaos there can arise a purified, and of course a severely modified, community. Violence must be suffered. A number of innocent victims must die so that the truly sinful will perish—though in this particular novel it is sinful "ideas" that are to be destroyed. The dying Stepan Verkhovensky, who has come to the realization that he has lied to himself all his life and has never known his own country, insists that the Gospel woman who has befriended him read the famous passage from Luke about the demons who flee from a man and enter swine—he is able, despite his feverish condition, or perhaps because of it, to underscore for us the apocalyptic moral of Dostoyevsky's novel:

> You see, it is just like our Russia. Those devils or demons coming out of the sick and entering into the swine—they are all the festering sores, all the poisonous vapors, all the filth, all the demons and the petty devils accumulated for centuries and centuries in our great, dear, sick Russia. . . . But the Great Idea and the Great Will pro-

> tects her from up above, just as it did that other madman possessed by demons; and all those demons, all that filth festering on the surface, will themselves beg to be allowed to enter the swine. Indeed, they may have entered them already! It's *us,* us and the others—my son Peter and those around him; and we'll hurl ourselves from the cliff into the sea, and I'll be the first perhaps, and all of us, mad and raving, will drown and it will serve us right because that's all we're fit for. But the sick man will recover and will sit at the feet of Jesus. . . .[2]

❧

Ritual violence has about it a preordained, even a supernatural quality. It is not necessary to experience it as *true,* to use Kafka's witty and despairing distinction, but only to experience it as *necessary.* Stavrogin must, like Prince Myshkin and Rogozhin, like Svidrigaylov as well, come to a violent end: even his innocent son, named for the passionate, warm-hearted Slavophil Shatov, must die. Liza, who has unwisely given herself to Stavrogin in a doomed attempt to "compress her life into one hour," has been contaminated by her love for him, and must die one of the ugliest deaths in the novel, at the hands of a rapacious mob. Powerful demons demand a powerful exorcism: obeying the instinct for tragic purgation, Dostoyevsky necessarily *sacrifices* his main characters as the novel comes to its tumultuous close. And it is a stroke of genius for Peter Verkhovensky to escape, for, acting as Dionysus, a psychopath in the guise of a "revolutionary," he is not human like the others and cannot share in their human fate. (Using a forged passport he very easily slips over the border—he is consequently loosed to the world, the diabolical progeny of Stepan Verkhovensky's "liberal blather.")

Much has been said of the unevenness of *The Possessed:* Dostoyevsky has been accused of creating caricatures rather than characters, and of exaggerating the imbecilic nature of his "anarchists." Several close readings of the novel have convinced me that this is not the case. Of course if *The Possessed*—like any of Dostoyevsky's work, beginning with *The Double*—is measured against the conventional standards of naturalism, it will seem somewhat feverish and improbable: but so will *King Lear* and *Hamlet*. Dostoyevsky uses brilliantly many of the devices of naturalism, the most obvious being his deliberately flat, blunt, reportorial style, which reaches peaks of eloquence only rarely (in certain stream-of-consciousness passages when Raskolnikov is delirious, for instance), and of course his reliance upon historical events (the famous Nechayev case of 1873)[3] but he is primarily a writer of myth. Or, more accurately, he is under the spell of a mythic imagination. The "horizontal" busyness of his longer works is part a consequence of his literary heritage, his conception of the nineteenth-century novel (specifically Dickens's) as melodrama serially presented, and part a consequence of his intuitive grasp of his material as dense, clotted, and recalcitrant. And mysterious: in fact unfathomable. *The Brothers Karamazov* is an unsettling novel to study because it appears to contain (consciously?—unconsciously?) its own double, a shadow or anti-novel that parodies the explicit novel's highly explicit concerns, and moves irresistibly toward tragedy. So much is affirmed—and rather noisily, too; yet so much is mocked and rejected. What precisely are we to believe? What are we to *feel?* In *The Possessed* the horizontal or linear concerns become vertical or thematic: the feverish complications of the plot symbolize the feverish complications of Russia as Dostoyevsky saw it, a great nation approaching Armageddon. (If he was not "realistic" in his portrayal of revolutionaries, he was at least prophetic.) Plot is transformed into theme, into symbolic meaning. Plot—and plotting itself—is metaphor. (For nearly everyone in the novel, innocent and murderous alike, is caught up in the activity of plotting.) The cacophony of voices, the wildly accelerated scenes translate into a philosophical, even a metaphysical statement, for there is a point at which "real life" must surrender to the forms of melodrama which do not distort it but in fact express it faithfully. It is presumably the narrator's own astonishment he expresses in such remarks, an astonishment that his quiet provincial town should so suddenly erupt into melodrama:

> It was a day of the unexpected; a day of the denouement of many plots and the beginning of many future intrigues; a day of sudden explanations and thickening mysteries. . . . It all ended in a way no one could have expected. Indeed it was a day of most extraordinary coincidences. [145]

That so prodigiously long and so luridly convoluted a novel as *The Possessed* evolves, nevertheless, with the structural coherence of a tragedy of Aeschylus or Euripides is a testament of Dostoyevsky's unparalleled genius. It has always been known that he is a marvelous creator of character—he is the equal of Dickens, and perhaps even the equal of Shakespeare, in this regard. But that he is a genius as a craftsman is perhaps less well known. It is, in fact, an embarrassing cliché of literary criticism that only short works of fiction, like novellas or short stories, exhibit perfect "form," and that any lengthy work inevitably suffers from a relative shapelessness. The naïve critic tries to compare *The Scarlet Letter* and *Moby-Dick*, discovering the one to be marvelously compact and the other sprawling and structurally unsound. But *Moby-Dick* is a masterpiece of structure, of a complexity that goes beyond anything Hawthorne would have dared to attempt; and it is to be presumed that the ordinary critic, infused with a myopic Jamesian sensibility, simply cannot *see* its vast magnificent form. My reading, over the years, of criticism on Dostoyevsky has led me to the conclusion that many of Dostoyevsky's critics are simply incapable of measuring his genius. Perhaps it is the case that the academic-trained critic will peer into a work of art in the hope of seeing his own reflection there, or certain "critical" qualities his professors in graduate school told him to admire—symmetry, unity of tone, precision, even brevity. Don't all literary works aspire to the condition of the well-wrought poem?

The "loose baggy monster" of Russian art is loose and baggy and monstrous only to the critic who confuses his own relative short-sightedness with an aesthetic principle. *The Great Gatsby* is a masterpiece of organization, but so is *The Brothers Karamazov. The Turn of the Screw* is deftly and beautifully orchestrated, but so is *The Possessed*—and *The Possessed* is an incomparably superior work. Yet Dostoyevsky is routinely accused of being slipshod and untidy[4]—critical clichés that cannot be honored if one studies his novels assiduously. (Why, one wonders, do people so readily assume that a large, ambitious work is necessarily any less subtle than a very short work? D. H. Lawrence's Ursula says, "A mouse isn't any more subtle than a lion, is it?")

Dostoyevsky had a lifelong interest in lancet architecture; he had even made a drawing of the cathedral at Cologne, and in later manuscripts he included extremely detailed drawings of granite portals, rose windows, and towers.[5] It seems quite probable that his fascination with vast, complex structures had something to do with the dense and multilayered art he created, in which the sub-text—the underpinnings—plays so powerful (though inarticulate) a role. There is a Gothic fussiness to his art, a lavish multiplication of detail, that allows the reader to imagine himself, temporarily at least, in a "real" world. (Hence the low-keyed comedy of much of the first section of *The Possessed,* where poor Stepan Verkhovensky and Mrs. Stavrogin attempt to make their way in liberal Petersburg circles. Dostoyevsky's Apocalypse will be firmly grounded in the domestic.) The novel is beautifully unified through the repetition of certain themes, primarily the theme of violation (of innocence, and of individual integrity); in Dostoyevsky's grandiose myth it is Nature itself, the very earth, that is betrayed and must be redeemed. Dostoyevsky's characters often ask one another passionately whether they believe in God. But what *is* God?—where does Dostoyevsky's God dwell? Not in the sky, surely, but in the earth: for, as Shatov says, a man who loses his links with his native land loses at the same time his his gods and his life's goals. The "possessed" characters of the novel are defined almost exclusively in terms of their relationship to their native land, which is God, a living, mystical, transcendental force, and it is significant that they are related to one another innumerable ways, some of them secret—quite as if they were all members of a single family. At the very center of the novel, however, are Stepan Verkhovensky and Stavrogin, who unify it by way of their very different passions. The novel begins with an aging deluded man who imagines himself "in exile," and ends with him in exile indeed, yet reunited with his nation and with God. It begins with a Russian prince in exile, a false Savior, and ends with his self-immolation, his ceremonial death. There are two distinct heroes—two tragic figures—and each undergoes a ritual degradation, a ritual suffering, and a ritual cleansing. That they are, in a sense, father and son, Old Russia and "New" Russia, gives the novel an allegorical depth that in no way detracts from its probability. For even tragedy, that "terrible sacrament of the god" of which Yeats speaks, can be transformed into the probable, the historically inevitable.

The Possessed, like the most ceremonial of Greek tragedies, is in essence a debate: a dialogue between characters in opposition who are at the same time curiously similar. If Stavrogin is announced in his mother's drawing room, it is Peter who appears. If Shatov speaks passionately and lengthily of the mystical force—the unquenchable will to reach an end and, at the same time, the denial of that end—that underlies all nations, it is really Stavrogin who speaks. ("I doubt that those are my exact words," he says cautiously.) Fedka, Stepan's former serf, is Peter's—and Stavrogin's—shadow, a creature with the doomed cunning of a Smerdyakov, the instrument of others' murderous wishes. Kirilov too echoes Stavrogin's ideas; and though Stavrogin rejects his incoherent praise of suicide, we learn from Shatov that Kirilov is Stavrogin's "creation"—or is Stavrogin, who also commits suicide, Kirilov's creation? Dostoyevsky's characters generally present mirror images of one another; it is perhaps too reductive to say that they are "doubles," but they certainly echo one another, and parody one another. (In *Crime and Punishment,* for instance, Raskolnikov's redemption through Sonia parallels Svidrigaylov's rejection by Dunya: the one lives, the other commits suicide.) Peter is Stavrogin's "ape"; he is a crook, and not a socialist, yet it is Stavrogin who has an "uncanny talent for crime." Stavrogin has a love affair with both Shatov's wife Mary and his sister Dasha, and as a consequence of a prank he is married to the demented Maria Lebyatkin; some years earlier, we are told, he raped and drove to her death a twelve-year-old girl named Matryosha. His tutor Stepan Verkhovensky was also the tutor of Shatov and Dasha, and Liza Drozdov; he is of course the father—the negligent father—of the psychopath Peter. So all are related. All mirror one another. "If the art of tragedy is to be defined in a single phrase," René Girard says, "we might do worse than call attention to one of its most characteristic traits: the opposition of symmetrical elements."[6]

It is in the enigmatic figure of Stavrogin that warring elements meet. Nearly everyone in the novel, male or female, looks to Stavrogin for guidance or inspiration or strength or emotional support; it is bitterly ironic that he is the means by which they are "possessed" while he himself remains untouched—in the words of Revelation which are read aloud by Tikhon in one episode (to Stavrogin) and by the Gospel woman in another (to Stepan Verkhovensky), he is neither hot nor cold but lukewarm. A demonic frenzy is loosed about him and through him, yet Stavrogin is dying of boredom. Like Raskolnikov in his cramped cell of a room, Stavrogin, though he wanders through Europe, though he makes a pilgrimage to Mount Athos, and visits Egypt, and even Iceland, goes nowhere at all: he is suffocating, doomed, trapped in the claustrophobia of a person who cannot love. As a personality he is as complex, and as fascinating, as Ivan Karamazov, who shares a number of his preoccupations and whose fate resembles his. He is a somewhat older version of Raskolnikov, a younger but equally jaded version of Svidrogaylov. As events accelerate at a dizzying pace about him he remains unmoved, untouched. His fatal ennui strikes us as being far more convincing than the passionate warmth of Shatov, or the mysticism of Kirilov: Stavrogin in his disjointed confession (which he does not trouble to make "artistic") articulates one of the tragic psychological problems of our time. How is one to escape boredom? How is one to apply one's strength—which is, in fact, "limitless"—to anything that is not an illusion, a simple pragmatic means of *applying oneself?* Revolutionary activities are certainly diverting, as are Slavophil beliefs; a marriage might, for a time, divert; then there are love affairs, card games, debaucheries, sin. Bouts of playful madness: leading an old gentleman by the nose, biting another old gentleman's ear. Giving away money. Taking part in idiotic social activities. One *ought* to commit suicide, to sweep oneself off the earth like a "pernicious insect," yet that calls for a generosity that is perhaps not forthcoming. It may be that the situation is more tragic than one can know, and that character itself is fate: for all that one *does* is a consequence of what one *is*. Even atheism is more admirable than worldly indifference, yet how is it possible for a person to remake his soul? It is said in the Book of the Apocalypse:

> These things saith the Amen, the faithful and true witness, the beginning of the creation of God; I know thy works, that thou are neither cold nor hot; I would thou wert cold or hot. So then because thou art lukewarm, and neither cold nor hot, I will spew thee out of my mouth. Because thou sayest, I am rich, and increased with goods and have need of nothing; and knowest not that thou art wretched, and miserable, and poor, and blind, and naked. . . .

Stavrogin is called many things by his disciples: he is a "wise serpent," a "magician," a "beautiful proud young god," a "fairy-tale prince," in Peter Verkhovensky's excited words, "who seeks nothing for himself but stays in hiding with a halo of sacrifice around his head." He is a "real gentleman," everyone decides after his refusal to kill a man in a duel (because he did not *feel* like killing anyone at the moment); "the most brilliant specimen of the young set." He is the "Sun." And yet he is a "demon of irony" who has led a *sarcastic* life in Petersburg and who is, perhaps, not altogether well. Kirilov, who loves him, says that he is caught up in a futile search for new sensations because he has become oversatiated with pleasure; Shatov, who also loves him, and who speaks to him with the blunt angry familiarity of a brother, accuses him of having married Maria Lebyatkin precisely because the senselessness and the disgrace of it bordered on "genius." Stavrogin, like Mitya Karamazov, does not content himself with teetering on the brink of the abyss—he plunges into it headfirst. Shatov says in disgust: "You married her to satisfy your passion for cruelty, your passion for remorse; you went through it for the mental sensuality. It was a deliberate laceration of the nerves" (241). Even so, Shatov confesses that he would be unable to prevent himself from kissing the ground Stavrogin has walked on. "I can't tear you out of my heart," he says.

He is to be the Savior of the Movement. Yet Lyamshin believes that he travels about "incognito," as a representative of "high government agencies." He is Peter Verkhovensky's "better half." He is Peter's "America"—which is to say, Peter's invention. As a child he was puny, pale, and strangely withdrawn, and far too emotionally caught up in his tutor's whims and moods. (Dostoyevsky hints of an unhealthy relationship between Stavrogin and Stepan Verkhovensky, but does not develop it; nor do the two men seem particularly aware of each other. Yet it is said that Stepan would wake Stavrogin up in the night and "pour out his wounded sensibilities," and the two would then sob in each other's arms. He had managed "to touch the deepest-seated chords in the boy's heart, causing the first, still undefined, sensation of the undying, sacred longing that a superior soul, having once tasted, will never exchange for vulgar satisfaction. . . . Anyway, it seemed a good idea finally to separate the teacher and his pupil, even though it was rather late" (42).)

Dostoyevsky's narrator is obviously ambivalent about him, and we may assume that Dostoyevsky himself did not know quite what to think. There is of course something morbidly fascinating and wonderfully romantic about the criminal, even the murderer—so long as the criminal is an attractive human being. Much is made of Stavrogin's appearance: if he were not so handsome Peter Verkhovensky would not have fallen so comically in love with him. Much is made, also, of his exquisite good manners; even the narrator must admit that Stavrogin is "the most elegant gentleman" he'd ever met. He is, astonishingly, both modest and bold. His physical beauty impresses everyone, and yet—

> This handsome head of black hair was somehow a little too black, his light eyes were perhaps too steady, his complexion too smooth and delicate, and his cheeks too rosy and healthy; his teeth were like pearls and his lips like coral. This sounds like a strikingly beautiful face, but in reality it was repulsive rather than beautiful. His face reminded some people of a mask. [44]

The Devil, perhaps? But a Devil with no clear perception of who he is and to what purpose even his cruelty might be directed.

We know that Stavrogin's self-loathing is a consequence of his violation of the child Matryosha, but we don't know, any more than Stavrogin does, what compelled him to that violation. The narrator speculates that he has never felt anger—even in moments of unbounded hatred he remains curiously aloof from his own passion. Though he is compared to the Decembrist Lunin, who deliberately sought danger throughout his life, he is really much more puzzling: his viciousness is cold and controlled "and, if it is possible to say so, *reasonable*—the most repulsive and dangerous variety there is." In the remarkable chapter "Stavrogin's Confession," which was censored and only published for the first time in 1922, Stavrogin sets forth the details of his Petersburg existence and presents what must have been at the time an astonishing document of "vice"—and an even more astonishing self-analysis of the sadistic-masochistic personality. Knowing himself despicable for having allowed a child to be beaten in his presence, because he had mislaid his own penknife, he nevertheless feels a pleasurable sensation "which burned me like hot iron and with which I became very much preoccupied." His more characteristic mood is one of appalling emptiness; he says that he could have hanged himself out of sheer boredom. Distractions are essential, and it hardly matters what they are:

> . . . I was then seriously preoccupied with theology. It distracted me a little but afterward things became even more boring. As to my political views, I just felt I'd have liked to put gunpowder under the four corners of the world and blow the whole thing sky-high—if it had only been worth the trouble. [418]

(This is the man who is Peter Verkhovensky's "better half.") Drawn to the child Matryosha, or perhaps only to the keen, disturbing sensations he feels in her presence, Stavrogin either rapes or seduces her, and afterward treats her with indifference. The intense excitement he has felt passes; he doesn't even hate her any longer; she too bores him. Quite deliberately he allows her to hang herself. And returns to his customary life of card-playing and debauchery, sensing himself both unaffected by the child's death and yet deeply and irrevocably changed. He formulates to himself the idea that he neither knows nor feels what evil is. "It wasn't simply that I had lost the feeling of good and evil, but that I felt there was no such thing as good and evil (I liked that); that it was all a convention; that I could be free of all convention; but that if I ever attained that freedom, I'd be lost" (426). He knows himself more thoroughly than Raskolnikov knows himself, for Raskolnikov acts under the delusion that to free oneself of conventional notions of good and evil would be to attain a kind of godliness, to be a superman. And Stavrogin is even more deadly than Svidrigaylov, who has also violated an innocent child, a deaf and dumb girl who hangs herself, because Svidrigaylov, for all his eloquent depravity, is still capable of a genuine feeling for Dunya; and at the very least he seems to have faith in physical sensation—in vice there is something permanent, he declares, "founded indeed upon nature and not dependent on fantasy, something present in the blood like an ever-burning ember, for ever setting one on fire and, maybe, not to be quickly extinguished, even with years."[7] Surely Svidrigaylov speaks —in part, at least—for his creator, just as the doomed Stavrogin does. Others may believe in fantasy—Raskolnikov, Shatov—but there is, at least, something undeniably *real* about the body's life. Yet even this philosophical hedonism strikes Stavrogin as empty; in the end he is more bored than ever.

Dostoyevsky's predilection for the dramatic image of the violated child can be accounted for on a fairly obvious level: of course he identified with his repulsive heroes, and must have felt a terrible attraction toward the violation, sexual or otherwise, of what he assumed to be pristine innocence. That the child might *not* be innocent is far more alarming than the fact of the violation itself: hence Svidrigaylov's suicide after his nightmare of the five-year-old harlot who opens her arms to him. (A passage of incomparable economy and power—far more moving than Stavrogin's somewhat similar experience.) In other works innocent women, often child-like, are misused, and from one point of view the entire plot of *The Brothers Karamazov* springs from old Karamazov's drunken rape of the feeble-minded Lizaveta. (It is from this union that Smerdyakov, the "epileptic chicken," the shadowy fourth brother who will be Ivan's instrument of revenge, is born.) Another feeble-minded Lizaveta is murdered almost by accident by Raskolnikov. Apart from the obvious interpretation of such events one might hypothesize that, in Dostoyevsky's imagination, the child or child-like woman is not female, primarily, but an image of Nature itself—innocent, near-mindless, possessing no language and very nearly no ego. In psychological terms the child is a symbol of the soul, and if a man "violates" it he is really violating himself; or he is dramatizing the fact that it has already been violated, or has been allowed to die. The degradation of the child not only necessitates suicide but *is* an act of suicide.

That this is a helpful reading of an admittedly obsessive concern of Dostoyevsky's is underscored by a consideration of Stavrogin's odd relationship with his wife Maria Lebyatkin, whom he has married in secret. It does not occur to anyone in the novel, and certainly not to Stavrogin, that he married her because he loved her—or saw in her qualities that might justify love. She is a fascinating character, one of Dostoyevsky's most inspired creations, and all the more remarkable in that one feels—as in the case of Smerdyakov too—that Dostoyevsky had no clear conception of the rich, provocative depths he was going to explore as soon as he allowed her to speak. Narrative in Dostoyevsky tends to be matter-of-fact, language tends to be prosaic, almost banal. Only when people are given voices or, in the case of the nameless Underground Man, have control of the narrative all along, do we experience the surprises that belong to a genuinely spontaneous art, an art that seems to be in the process of being born as we attend. Smerdyakov is dismissed as an ignorant inconsequential lackey by the narrator of *The Brothers Karamazov,* yet his subtle intelligence rivals Ivan's; and Maria Lebyatkin, though ostensibly deranged, exhibits an independent and even rather daring intelligence that compares favorably with that of the characters who surround her. It is a sign of Dostoyevsky's genius that Maria is both legendary and "real": a figure out of myth, a feminine component of Stavrogin's despairing (because too intellectual) masculinity, and yet a very convincing, very credible person. She is perhaps an image of Stavrogin's soul, and at the same time an image of the violated and betrayed soul of Russia. In which case Stavrogin, like a number of Dostoyevsky's criminal heroes, takes on a Christly role—but he is a broken, cynical Christ, a Savior no longer capable of redeeming others, or himself, let alone his nation. When he comes to visit Maria with the intention of helping her she is, significantly, asleep or in a trance, experiencing a vision of

Stavrogin himself that suggests that he does not really want to help her: he wants, not quite consciously, to kill her.

She sees him as the Devil, though she wants to see him as her Prince, her falcon, her Savior. What worries her is that she "may fall very much out of love" with him. He angrily rejects her title of Prince and she lapses into a demented but shrewd speech accusing him of being a poor actor impersonating Stavrogin. "Or have you killed him?" she asks. "Admit it!" Sensing his desire to murder her before he has become aware of it himself, Maria shouts that she is not afraid of his knife: and in a surprising act of violence Stavrogin (who would not defend himself against Shatov's public blow) pushes her against a settee so hard that her head and shoulders strike painfully. He runs out accompanied by her curses: "You false pretender!"

Nature, the earth, Russia, the "eternal feminine": Maria must suffer her husband's betrayal, and even his murderousness. She is, like Lizaveta of *Crime and Punishment,* and like Sonia to some extent, a "perfect victim." She is a virgin troubled by despairing dreams of a dead baby, sunken in a pond. She is a kind of oracle, with her deck of worn fortune-telling cards and her little hand-mirror, a dreamer, with whom Shatov can talk quite intimately: Shatov being in a more reverent relationship with the mystical impulses she embodies. She tells him of a conversation she had with a lay sister in a convent about the Mother of God, the "great mother earth" itself, and of her worshipping of the earth—

> I climbed the mountain, faced the east, and kissed the ground; and I cried and cried and didn't even know how long I stayed there crying or what was happening around me. Then I got up and turned around and watched the sun setting. . . . [140]

(It is significant that Raskolnikov is told by Sonia to kiss the earth which, in murdering the pawnbroker and her sister, he has defiled; and Shatov shouts at Stavrogin much the same advice.) Bishop Tikhon speaks of the torture of those who have torn themselves away from their own soil, and clearly foresees Stavrogin's suicide. The allegory at the heart of *The Possessed* seems to turn on the "fall" of man into rational, conscious, "masculine" knowledge (symbolized here, of course, by those evil ideas from the West—socialism, anarchism, perhaps even democracy), and the necessity of suffering and atonement. What must be broken is the proud, egocentric will: the sinner either goes insane, like Ivan Karamazov, or kills himself, like Stavrogin, or experiences a conversion of the personality so complete that it has the quality of a miracle. (So Stepan Verkhovensky, the novel's other tragic figure, is converted on his deathbed to a faith in Russia and in love that resembles Shatov's.)

Critics who identify Dostoyevsky with his Slavophil characters are certainly misreading him, and it seems quite simply an error—a very popular error, to be sure—to insist upon using passages from his letters or journal in order to "explain" *The Possessed.* It is true that Dostoyevsky in writing to a friend spoke of the novel-in-progress as "tendentious"; he went so far as to say that he was so anxious to express certain ideas that he didn't care if they ruined his novel as art. "Let it turn out to be only a pamphlet, but I shall say everything to the last word," he told his friend Strakhov on April 5, 1870. But a writer's remarks can never be taken as serious evidence. They are remarks—nothing more. *The Possessed,* seven hundred dense pages long, is hardly a pamphlet, and the ideas dramatized in it are hardly simple ones. Dostoyevsky is clearly more sympathetic with certain characters than with others, and yet no one, not even Shatov, is spared his irony. Shatov's beliefs are admirable, particularly in the context of this demonic work, and yet—are they not subjected to Stavrogin's chilly cynicism in order that Dostoyevsky can put them to the test? And if they are fairly reasonable beliefs are they not (as Shatov's fate suggests) utterly quixotic?

> Socialism is godless—it proclaimed in its very first statement that it aims at an organization that does not presuppose God; that is, an organization based on the principles of reason and science exclusively. But reason and science have always performed, and still perform, only an auxiliary function in the life of peoples, and it will be like that till the end of time. Nations are formed and moved by some other force whose origin is unknown and unaccountable. . . . It is the force of an incessant and unwavering affirmation of life and a denial of death. It is the spirit of life, "river of water of life" as the Scriptures call it, the drying up of which is threatened in the Apocalypse. . . . I call it simply the search for God. The objective of any nationalist movement in any people at any time is actually a search for God, for their own, national God—and it must be, above all their own God. . . . God's personality is a synthesis of the entire nation from the beginning of its existence to its end. [237]

These are ideas first expressed by Stavrogin, some years before, and now rejected by him. Or perhaps they merely embarrass him. And what are we to think of Shatov's advice (which will be echoed by Chekhov in *The Cherry Orchard*) "Listen, Stavrogin, find God through labor. That is the essence of everything. . . . Find God through labor!"

Which is to say: find God by a denial of your own intelligence.

It is Peter Verkhovensky who strikes the reader as Stavrogin's most conspicuous, and most compelling, double. He is an extraordinary creature, closer perhaps to mythology than to nineteenth-century Russia, though contemporary terrorism is probably fueled by an amoral zest for action, for death, for *exercise,* that resembles Peter's. It is too simple to call him a psychopath, though of course he feels no remorse for anything he does, and seems incapable of even remembering it for very long. On his way to murder Shatov he stops in a restaurant and devours a steak; he eats poor Kirilov's chicken dinner on the night Kirilov has promised to commit suicide. His manner is noisy, blustering, shrewd, and even—oddly—rather charming. When he appears the narrative springs ahead: he has the outrageous verbal energy of a Marmeladov, of an Underground Man. Flippant, breezy, knowing, cynical, good-natured, he is at the same time capable of gratuitous sadistic acts that make Stavrogin's seem, by contrast, almost abstract or visionary. Yet he *is* sane in a way others are not. His amazing egocentricity makes him almost innocent, almost modest: Stavrogin remarks of him that he passes himself off as "only an agent" when in fact the society to which he belongs consists of no one except himself. If he is a freakish product of modern Western thought he is, even more significantly, the cast-off son of Stepan Verkhovensky, who has toyed casually with ideas all his life and has never considered what the consequences of his ideas might be. He hopes to drive his father to despair—but it is nothing personal, simply part of the plot. With his wide complacent maddening grin he rushes about everywhere, as his listless Fairy-Tale Prince lies half-dozing on a couch; he so charms the silly wife of the provincial governor, by a method of flattery so gross that it cannot fail to work, that he practically eats and sleeps in the Governor's mansion. He is, or is not, really in the pay of the Secret Police. He knows everything: he is literally everywhere, like a god or a demon: if Tolkachenko, one of the "anarchists," whispers something into the ear of another man, Peter overhears it; if Liputin pinches his wife black and blue in the secrecy of his bedroom, Peter knows. The devilish Fedka himself will not escape Peter. It is suggested that he has plans to murder even Stavrogin, whom he idolizes; though he is Stavrogin's ape he nevertheless feels that since he "invented" Stavrogin he should control him.

Peter's ideas, released in a feverish torrent, foreshadow the coolly Machiavellian, and altogether too reasonable, ideas of Ivan Karamazov's Grand Inquisitor. Though Dostoyevsky was certainly a reactionary politically[8] it is absurd to claim, as his translator David Magarshack does, that he cannot be taken seriously as a political novelist. Are Peter Verkhovensky's terrorist ideas really so improbable?—they are apt to strike us today as prophetic. He speaks, for instance, with approval, of Shigalov's scheme for the transformation of society ("I started out with the idea of unrestricted freedom and ended with the idea of unrestricted despotism"), which will necessitate a complex spy system in which everyone spies on everyone else. He explains disjointedly, hurrying after his hero Stavrogin who wants only to escape him:

> Each belongs to all and all belong to each. All are slaves and equal in their slavery. In the first place, there is a lowering of the level of education, science, and arts. The highest level . . . is accessible only to those with the greatest abilities. . . . The most gifted men cannot help being tyrants and they have always perverted others more than they have been useful to them: so they must be ostracized or put to death. . . . But in the herd there is equality. . . .
>
> All we have to do is organize obedience—that's the weak point in this world of ours. . . . Everything must be reduced to the common denominator of complete equality. . . . But they need to be shaken up too and we, the rulers, will take care of that. Because slaves must always have rulers. There'll be total obedience and total depersonalization. . . . [399]

Peter declares that he is a crook, not a socialist; he is a nihilist, but one who loves aristocracy and beauty. He strikes the reader as half-mad because of his inarticulate love for Stavrogin as a man, rather than as a symbol, but Dostoyevsky does not explore this—indeed, one has the impression that he is not consciously aware of Peter's erotic excitement at all. Peter is Stavrogin's ape and shadow and brother but he is also his lover, and though he even kisses Stavrogin's hand the gesture seems to be intended by Dostoyevsky as merely symbolic. Yet it is far more:

> "Stavrogin, you're an extraordinarily handsome man!" Peter cried in what might have been rapture. "Do you have any idea how handsome you are? . . . Ah, I have made a thorough study of you: I've often observed you when you weren't looking. . . . I love beauty. I'm a nihilist, but why shouldn't I love beauty? As if nihilists couldn't love beautiful things! Nihilists can't love idols, though, but me—I love idols and you're my idol! . . . You are the leader, the sun, and I'm your worm—" [400]

With Stavrogin he is not acting a part: he speaks quite sincerely, though feverishly. Nothing is more difficult, he has admitted, than being oneself.

The last we see of Peter, after the brutal murder of Shatov, he is making plans to join several wealthy landowners at cards in the first-class compartment of the Petersburg train. He simply disappears: unlike the anarchist Nechayev on whom he is modeled, he is never caught. *The Possessed* is the most disturbing of Dostoyevsky's novels because the guilty are not all punished, nor do they punish themselves. Evil, chaos, the demonic itself, is loosed into the world. Beyond the formal tragedy of Stavrogin and Stepan and the community there is a bitterly ironic, comically grotesque universe, or void, in which the concept of justice does not exist: energy like Peter's is a phenomenon as amoral as the weather, beyond all provincial notions of good and evil. This is the "lost" state of the soul of which Stavrogin speaks in his confession. It is the primeval chaos, the formlessness out of which humanity was shaped; no ritual, no ceremony, can control it. Hence Peter's effortless escape. And Dostoyevsky's implied warning for the future. "I think I should inform you, Mother," Stavrogin says early in the novel, "that Peter Verkhovensky is some sort of a universal peacemaker; peacemaking is his role, his forte, his disease" (186).

❧

As society approaches crisis and breakdown, preparatory to a reaffirmation of its identity, it provokes private disintegration, private ceremonial exorcism. To what extent the solitary suicide knows himself a participant in a vast communal phenomenon one cannot guess: to sense that the ground is slipping out from under everyone's feet might be salvation to one individual, but a confirmation of despair to another. Contemporary society—by which I mean primarily ours in America—seems curiously able to accommodate itself to the loss, year by year, of an astonishing number of persons of all ages to suicide and suicidal self-destruction. And there are thousands, perhaps millions, who, in plunging into the anti-intellectual and in many cases delusional systems of various religious cults (Hare Krishna, the Unification Church of the Reverend Moon, Divine Light, etc.), have opted for a less evident sort of suicide. Dostoyevsky's Russia was in the process of an even more violent transition: a tragic rejection, as Dostoyevsky saw it, not only of the Church and the established order, but of God—and all that God symbolizes in the human soul. That Dostoyevsky preaches so extreme a doctrine, that it never seems to have occurred to him that socialism, founded on "reason and science" (which he abhorred despite his own background as an engineering student) might one day accommodate itself to the religious instinct in the people, is unfortunate, but in a way prophetic; and in any case it inspired him to an apocalyptic fervor that resulted in extraordinary creative activity. We know that terror of change—any change—is a characteristic of the primitive mind, but that primitive mind is always with us, and has the power at times to create an uncannily beautiful poetry of despair: *The center cannot hold, mere anarchy is loosed upon the world,* we have been moved by our greatest poets' angry admonitions over the centuries, even when we have felt ourselves unable to sympathize with their politics:

> The specialty of rule hath been neglected;
> And look, how many Grecian tents do stand
> Hollow upon this plain, so many hollow factions.
> When that the general is not like the hive
> To whom the foragers shall all repair,
> What honey is expected? Degree being vizarded,
> Th' unworthiest shows as fairly in the mask.
> The heavens themselves, the planets, and this centre
> Observe degree, priority, and place,
> Insisture, course, proportion, season, form,
> Office, and custom, in all line of order;
> . . .
> But when the planets
> In evil mixture to disorder wander,
> What plagues, and what portents, what mutiny,
> What raging of the sea, shaking of earth,
> Commotion in the winds, frights, changes, horrors,
> Divert and crack, rend and deracinate
> The unity and married calm of states
> Quite from their fixure! O, when degree is shak'd,
> Which is the ladder to all high designs,
> The enterprise is sick.
> . . .
> Take but degree away, untune that string,
> And hark what discord follows. Each thing meets
> In mere oppugnancy: the bounded waters
> Should lift their bosoms higher than the shores
> And make a sop of all this solid globe;
> Strength should be lord of imbecility,
> And the rude son should strike his father dead;
> Force should be right, or rather, right and wrong—

> Between whose endless jar justice resides—
> Should lose their names, and so should justice too.
> Then every thing includes itself in power,
> Power into will, will into appetite,
> And appetite, an universal wolf,
> So doubly seconded with will and power,
> Must make perforce an universal prey
> And last eat up himself.[9]

(The very spirit of *The Possessed.*)

It is Kirilov, the only shadow-self of Stavrogin to commit suicide, who has been making a study of the increasing incidence of suicides in Russia. Like most of the principal characters he has been abroad for several years, and has avoided people; it is thought that he has become somewhat alienated from his homeland. The narrator considers him insane, but Stavrogin takes him seriously, and his preoccupation with God—"God has tormented me all my life," he says —seems to be Dostoyevsky's own. In brief, Kirilov thinks of himself as a redeemer: his will be the first suicide in history to be committed for no purpose other than that of establishing man's free will (and the non-existence of the "old" God). There will be new era, a new humanity. History will be divided into two parts: from the gorilla to the destruction of God and from the destruction of God to the physical transformation of man and the earth. Man is to be a god—even physically. Kirilov preaches to the impatient Peter Verkhovensky, who is waiting for him to kill himself:

> I must affirm my unbelief, for there's nothing higher for me than the thought that there's no God. The history of mankind is on my side. Man kept inventing God in order to live, so as not to have to kill himself. To this day, the history of mankind consists of just that. I am the first man in history to refuse to invent God. I want it to be known always.
>
> . . .

> I'll be the first and last, and that will open the door. And I'll save them. That alone can save people, and the next generation will be transformed physically. Because the more I've thought about it, the more I've become convinced that, with his present physical make-up, man can never manage without the old God. [636–37]

Yet Kirilov is a reluctant god. We see him infrequently, but on two occasions he is playing with a ball; in one instance he is caught up in a game with an eighteen-month-old child and appears to be enjoying himself immensely, in the other he is doing exercises. By temperament he is not an unhappy man. He tells Stavrogin and Shatov that he is, in fact, very happy, as a consequence of experiencing certain "eternal moments" that obliterate all disharmony. Like Prince Myshkin (and Dostoyevsky himself), Kirilov is subject to ineffable "mystical" visions of the sort that sometimes precede an epileptic fit, though not necessarily; these visions have the power to reorganize the personality and to cleanse perception in such a way that the individual's ego is destroyed. Kirilov explains:

> There are seconds—they come five or six at a time—when you suddenly feel the presence of eternal harmony in all its perfection. It's not of this earth; I don't mean by that that it's something heavenly but only that man, as he is constituted on earth, can't endure it. He must be either physically transformed or die. It is a clear, unmistakable sensation. It is as though you were suddenly in contact with the whole of nature, and you say, "Yes, this is the truth." When God was creating the world, He said, after each day's creation, "Yes, this is the truth; it's Good." It's not elation, really, it's simply joy. . . . If it lasted for more than five seconds, the soul wouldn't be able to stand it; it would have to disappear. [609]

To Stavrogin, whose own death is growing inside him, whose own role in the Dionysian frenzy is slowly becoming evident, Kirilov speaks with a brotherly frankness; one feels that Kirilov is the person to whom Stavrogin feels closest, though as always he is circumspect and guarded. Ivan to Kirilov's Alyosha: watchful, puzzled, envious. "Have you ever looked at a leaf, the leaf of a tree?" Kirilov asks him. "I saw one recently. It was yellow, with some green in it and a bit wilted at the edges. The wind was blowing it around. When I was ten, in the winter, I'd close my eyes and imagine a leaf—green, bright, with those little veins, and glistening in the sun. When I opened my eyes, I couldn't believe it possible because it was too beautiful. Then I'd close my eyes again."

"What is that," Stavrogin asks, "an allegory?"

"N-no, why? I meant no allegory, just a leaf. One leaf. It's beautiful, a leaf. Everything is good."

I meant no allegory, just a leaf. So Kirilov in his mystical state not only knows himself happy, but believes that everyone is happy—man is unhappy, in fact, because he doesn't realize he is happy. The vision is Father Zossima's as well: Everything is good. And if Stavrogin believes in God he doesn't *believe* that he believes.

So Kirilov is a reluctant god, a reluctant sacrificial victim. Yet he must kill himself in order to affirm his free will, and the free will of all of humanity. Personal contentment is not an issue. A love of children, of nature, of life itself—all this is irrelevant. Kirilov kills himself to escape the fear of death as well (a quite logical motive shared, no doubt, by most suicides), and because, in the era in which he lives, there is no God and God's absence cannot be endured. His gesture of free will can be interpreted, ironically, as a response to the malaise of his society as direct, and as helpless, as the various responses made by Stepan Verkhovensky and Mrs. Stavrogin and Governor von Lembke and his wife and their circle, nearly all of which are comic. It is his tragic condition not to see that his attempt to affirm his freedom through suicide is an attempt to validate a principle that, in another era, in another Russia, he would not need to make. That his "affirmation" is utterly futile is borne out not only by the fact that, caught up in the stampede of events, no one even thinks about him, but also by the comically grotesque nature of his suicide itself. (A brilliant scene, one of Dostoyevsky's most inspired moments. Though Nabokov routinely ridiculed Dostoyevsky it is obvious that he drew upon the terrifying and jocose spirit of this sequence, and similar sequences in Dostoyevsky's work, for the death-scenes of Clare Quilty and John Shade.) Why does Kirilov *really* commit suicide? For a reason Dostoyevsky could well comprehend:

> Once three crosses stood in the center of the earth. One of the three of the crosses believed so completely that He said to another: "Today you will be with me in paradise." By the end of the day, both of them died. They went and they found—no paradise, no resurrection—nothing. His words didn't come true. Listen now—that Man was the best on earth—He represented that which makes life worth living. The whole planet with everything on it is sheer insanity without that Man. There hadn't been any-one like Him before nor has there been since—never; and therein lies the miracle—that there never has been and never will be such a Man. Now, since the laws of nature didn't spare even Him, didn't spare even that miracle, and forced even Him to live among lies and to die for a lie—it proves that the whole planet is based on a lie and an inane smirk. It proves too that the laws of nature are a pack of lies and a diabolical farce. So what's the point of living? [637]

(According to Dostoyevsky's wife, Holbein's painting "Christ Taken from the Cross"—as a rotting corpse—made a powerful impression on him; he remained standing in front of it for a considerable period of time. As described in the Grossmann biography, this particular scene—Dostoyevsky in deep, troubled contemplation before the Holbein painting—is extremely moving. One has a vivid sense of Dostoyevsky as a *participant* in his own tragic fiction.)

Irving Howe remarks of *The Possessed* that it is "drenched in buffoonery."[10] It is certainly a comic work, intermittently; tragic in scope and in seriousness, yet riddled throughout with a savage comic and satiric spirit. I find incomprehensible various critics' charges that the work is "uneven" and that the rapid shifts in tone and mood are flaws. Certainly they are deliberate, as similar irreverent passages are deliberate in Shakespeare's greatest tragedies. One can argue that there are comic-grotesque elements in certain Greek tragedies, even, and that the form, far from being damaged by such "shifts," is powerfully enhanced by them—when, of course, they are skillfully executed. Kirilov's bout of madness at the very end of his life, his biting of the terrified Peter Verkhovensky's finger, are devastating strokes, yet absolutely necessary. An isolated tragic gesture *is* swallowed up in the comic indifference of life, of things in motion. And the more gently comic aspects of Stepan Verkhovensky's last hours, for instance, are absolutely just. One cannot imagine *The Possessed* without them.

As catastrophe approaches and "degree" is lost, anything at all can happen. King Pentheus who has scorned women finds himself dressed in women's clothing, stuck at the very top of a tree; Lear, whose comfort has meant so much to his dignity, runs wild on the heath and is quick to content himself with the smallest gesture of solicitude. It is quite possible that Stavrogin kills himself partly because—as Bishop Tikhon senses, with characteristic shrewdness—he will not be able to bear ridicule. Governor von Lembke is destroyed by the lively demons that have sprung out of the earth to torment him: we learn that he is recovering from his nervous breakdown in a place far from home. No doubt it is the case that Dostoyevsky's portrayal of Turgenev as the absurd Karmazinov is grossly unfair, but it is funny, and the "great author's" pomposity and his humiliation before the crowd are quite necessary in the novel—just so is fame, and even "greatness" and "genius" itself, brought low. Indeed, the literary fête at the Governor's mansion, arranged by the silly, vain, and utterly believable Mrs. von Lembke, is one of the best sustained comic sequences in all of literature. One cannot imagine Dostoyevsky's ostensible master, Dickens, doing better. (The appearance of Captain Lebyatkin on stage is outrageous enough, but the appearance—at the very end, after both Karmazinov and Stepan Verkhovensky have been routed—of the maniac who punctuates his furious speech with a tic-like waving of his fist, and who finds the din of the gathering stimulating, is a brilliant stroke.)

It is Stepan Verkhovensky, however, who suffers most. His pride in his social pretensions has already been sadly undermined; now he submits to the crowd's jeering not only his severely modified political beliefs but his very self, the essence of his being. As the crowd grows more and more restless, and hecklers shout, he pleads with them as "a heart-broken and rejected father" to hear him out. As if the "literary quadrille" had not degenerated into a farce, he addresses them desperately in a squeaking voice:

> . . . Shakespeare and Raphael are of greater value than the emancipation of the serfs, than nationalism, than socialism, than the younger generation, than chemistry—and perhaps even than mankind itself! And it is this way because they represent the very highest human achievement, an achievement of beauty without which I wouldn't be able to go on living. . . .
>
> Let me tell you that mankind could survive without the

> English, without the Germans, and most certainly without the Russians; that it could subsist without science and even without bread. But it is impossible to do without beauty because then there would be nothing left for us to do in the world! [504]

He bursts into hysterical sobs. An angry divinity student accuses him of having sold his serf Fedka into the army, and being therefore responsible for Fedka's crimes, and he shouts his curse at the audience and runs offstage, to be followed by a genuine maniac.

The long sequence in which Stepan goes on foot to search for Russia is a masterpiece of sentiment checked by comedy. The poor old man is obviously deranged, and it soon becomes clear that he is dying, but the incoherent yet oddly perceptive monologue that attends his pilgrimage is enlivened by a sense—if not of comedy, perhaps of simple geniality. "This man was very dear to me," the narrator has thought earlier, and it seems clear that Dostoyevsky, who has been satirizing him for hundreds of pages, shares in this sentiment. In a state of collapse Stepan sees his former life as a tissue of lies. His judgment is not overly extreme; his life *has* been lies. Not only that, he is lying still. "The main trouble," he says, "is that I believe myself even while I'm lying. The most difficult thing in life is to live without lying and—and not to believe in one's own lies" (668). Mrs. Stavrogin's overbearing manner adds to the comedy; the Gospel woman repeats Stepan's confused, self-glorifying tale of having been loved by two women, one of them "such a fatty" (a slip of the tongue for "beauty"), and Stepan's deathbed announcement that he has always been in love with Mrs. Stavrogin is badly qualified. Feverish, dying, he attains a rapturous vision reminiscent of Kirilov's, and though it is exceedingly difficult to know how Dostoyevsky wishes us to interpret it, the scene itself is a memorable one; indeed, one has the feeling that the much-derided figure of Stepan Verkhovensky is the novel's central concern. The death of an old god, even if the old god has been somewhat ridiculous, is nevertheless tragic. One feels his death with as much emotion as the horrible death of Shatov, perhaps because Dostoyevsky has given him so unique a voice—

> Friends, I need God if only because He is the only Being who is capable of loving eternally. . . . My immortality is necessary if only because God would not wish to do anything unjust and put out the flame of love once it was kindled in my heart. And what is more precious than love? Love is higher than existence. . . . Since I have come to love Him and am happy because of this love, how could He extinguish me and my happiness and turn me into a zero? If God exists, then I, too, am immortal! *Voilà ma profession de foi*— [678–79]

Pretentious to the very end, old "liberal" Russia cannot resist a final French exclamation.

With Stepan Verkhovensky's death, and Stavrogin's suicide, the Walpurgisnacht concludes. The entire account is being given some three or four months later, by the unimaginative but presumably reliable Govorov, an anonymous Everyman, a survivor, a Russian whom the demons were not able to possess. *He*, and not those brilliant others, is Dostoyevsky's future.

Notes

1. *Crime and Punishment,* trans. Constance Garnett (New York, Bantam edition, 1962), pp. 469–70.

2. *The Possessed,* trans. Andrew R. MacAndrew (New York, Signet edition, 1962), p. 671. All quotations from *The Possessed* in this essay are from this edition, and subsequent page numbers will be given in brackets in the text.

3. Sergy Nechayev, a twenty-two-year-old disciple of Bakunin and a former divinity student, organized student disturbances at the Petersburg University in 1868 and 1869. After having organized a "Society of National Retribution" in Moscow, as a part of the World Revolutionary Movement, he and four members of a "group of five" murdered the fifth member in a way that closely parallels Shatov's death at the hands of the Five. The case was of course a sensational one that interested Dostoyevsky greatly; according to David Magarshack, who did a translation of *The Possessed* (under the title *The Devils*) for Penguin Books, Dostoyevsky even found the model for Kirilov among Nechayev's followers. (See the Translator's Introduction, p. xi, in the 1976 edition.)

4. By, among others, Magarshack, who prefaces his translation with an extraordinarily mean-spirited piece on his subject. It is a curious custom, this publication of, in paperback editions, self-important prefaces that in many cases distort or misread the text that follows.

5. Leonid Grossmann, *Dostoyevsky*, trans. Mary Mackler (New York, 1975), pp. 44–45.

6. René Girard, *Violence and the Sacred*, trans. Patrick Gregory (Baltimore, 1977), p. 44.

7. *Crime and Punishment*, p. 405.

8. But Dostoyevsky's fascination with the figure of the self-sacrificing revolutionary (in contrast to Peter Verkhovensky, for instance) would seem to suggest that he was divided on the subject. It is known, for instance, that he planned to write a sequel to *The Brothers Karamazov* (the principal novel, in fact—the lengthy one he published was only an "introduction") in which, twenty years later, his hero Alyosha would leave his monastery cell and become a revolutionary and die for his idealism. He "would have searched for the truth and in his quest would naturally have become a revolutionary," Dostoyevsky said. See Grossmann, p. 587.

9. Shakespeare, *Troilus and Cressida*, I, iii, 78–88, 94–103, 109–24. The quote is given at such length both because it so brilliantly gives voice to Dostoyevsky's sentiments and because, in the briefer form in which it is usually given, it does not adequately suggest Ulysses' passion.

10. Irving Howe, "Dostoyevsky: The Politics of Salvation," in *Politics and the Novel* (New York, 1957).

"Is This the Promised End?": The Tragedy of King Lear

Thou art a soul in bliss; but I am bound
Upon a wheel of fire, that mine own tears
Do scald like molten lead.

Lear

The moment of Lear's awakening is one of the most moving scenes in our literature, coming as it does after so much grotesque and senseless horror; it marks not simply the reconciliation of King and mistreated, exiled daughter, the reconciliation of the tyrannical, aggressive Lear and his loving, all-forgiving Cordelia, but the mysterious moment of "awakening" of the soul itself—for Cordelia, with her unearned kiss, symbolizes that moment of grace that forces the tragic action to a temporary halt, and allows a magical synthesis of the bliss of eternity and the tragedy of time that is so powerful in Shakespeare, because it is so rare.

It is moving, yes, but bitterly moving, and our emotions will be turned against us shortly, for the visionary experience of a timeless love cannot compete in Shakespeare with the tragic vision, the grim necessity of history. Only when he chose to call attention to the magical—and therefore "unserious"—elements of his own art-work, as in *The Tempest*, could Shakespeare go beyond the terrible tradition of history, that enemies be put to death, that no one be forgiven except the dead. In reality, history cannot be stopped, and history is no more than the recording of men's actions against one another—so Shakespeare might have concurred with Napoleon's cynical remark that history is the only true philosophy, and he would have eagerly chosen as a villain the man of modern times who, like Edmund, placed so passionate a faith in his ego's powers as to claim that such sentimental concepts as "friend" and "enemy" do not exist except as the ego forces them into being.[1] We accept, unquestioning, the prejudice of a personality that disguises its pessimism in the form of art, especially if the art is that of "tragedy"—which demonstrates by its surface action the rightness of such a prejudice, but only by its surface action. The mysterious core of tragedy is its ritualistic affirmation of the life-force; as a form of religious observation, tragedy becomes "artistic" only as the artist steps forward to declare his individuality, his unique powers of perception. No one has really written about tragedy from the inside—that is, from the point of view of the writer of tragedy, who deals not only at second-hand with the spectacle before him (or at third- or fourth-hand, since as late as the time of Pope true genius was "carefully, patiently, and understandingly to combine," not to invent), but immediately and intimately with his own personality, his largely unconscious attitudes grouped as external elements of character, event, in *King Lear* even as setting. If the Shakespeare who brought together the various lively elements that constitute *Hamlet* could have anticipated, or imagined, the naïve response of a Partridge (in *Tom Jones*) to that work, he might have had faith that, for some members of his audience, or for some layers of the human personality, the original magic of the ritual still worked. Yet it seems to me doubtful that Shakespeare did believe this: moments of transcendence in his plays are usually fleeting, often expressed by women, and in any case when they are brought to trial against the "cheerless, dark, and deadly" night of the unredeemed universe, they are always defeated. External history takes precedence over subjective experience, and the violent wheels that are individuals, mad for power, must turn full circle; whatever "promised end" the soul yearns for, imagining that a certain measure of suffering has crucified its sinful egotism, must be thwarted by the demands of history, which is unredeemed.

For most writers, the act of writing is itself a triumph, an affirmation, and the anguish experienced by an audience is not really in response to an emotion within the work itself (since real life would furnish much more convincing emotions) but the artist's genius, his ability to transmute into formal images an archetypal human drama. In the case of tragedy, this is an inconsolable grief that nevertheless testifies to a higher, supreme order—not the raw ritual any longer, which is experienced immediately as "religious" and not enjoyed in our sense of the word, but the ritual brought into human terms, incarnated into flesh, into heaving, bleeding conflict.

"TRAGIC" VISIONS

> *Why is the underplot of King Lear in which Edmund figures lifted out of Sidney's Arcadia and spatchcocked onto a Celtic legend older than history?*
>
> *Stephen Dedalus*
> *in the library scene of* Ulysses

As tragedy evolves from simple ritual into art, and into increasingly complex, stylized, and individualized art, a new force enters history—the diminishment of tragic "elevation" in the anonymous, rather democratic art of folk-tales and ballads, which always remain for all the wisdom they convey more or less artistically naïve; and, in formal art, the increasingly important factor of the self-conscious and self-declaiming creator, the arranger of the elements of ritual. A deliberate and deliberating consciousness asserts itself. When scholars like Hardin Craig, G. B. Harrison, and Russell Fraser draw our attention to the discrepancy between the Lear sources and Shakespeare's transformation of them[2] —as well as to the violent yoking-together of the Lear and Gloucester stories, never before united—we must remember that individual expressions of the tragic vision of life, however aesthetically and emotionally powerful they appear, are, first of all, to the artist a challenge of his individual artistry and an opportunity for him to experiment with partly conscious or totally unconscious elements in his own personality; but only in so far as these liberated elements can compete with the principle of reality itself, in tragic times usually represented by—not symbolized by—a political and social order involving a great deal of oppression. What we experience as infinite and universal, then, must be seen as a direct response to a given environment: not necessarily our environment, but valuable so far as the repressive nature of any force external to the individual can be externalized as a historical given. Is the tragic view of life necessarily the highest view of life, or the most beautifully rendered view of any life possible at the time of its having been rendered?—which is a way of questioning our usual acceptance of the artist's "formal" message (which the environment of his time forced into him and then from him) to the exclusion of those incontestably exciting moments, at times no more than in the interstices of the overwhelming general action, in which the liberating forces, the rebellious forces of life itself, are honored.

Harry Levin states bluntly that he can see "very little point in pretending, through some Hegelian exercise in cosmic optimism, that tragedy is other than pessimistic,"[3] and yet it seems possible that one can redefine the concept of "pessimism" itself and determine whether, in certain historically determined works of art, there is not a possibility of some transcendence, however forced by the conventional plot to be defeated. Not that Desdemona, Cordelia, Edmund, Hotspur, Falstaff, and others who cannot be contained within the established society are defeated—but that they have been imagined into being at all, that their voices, their imprudence and vitality, have been given any expression whatsoever—this does represent a triumph of the artist's personality, and we have only to remove the troublesome rebels from these works to see how pointless, how nakedly propagandistic, the "tragic vision" would have been. And how inexpressive of the complexity of Shakespeare's genius! But if this does not quite answer the charge that tragedy fails to elevate, that it is profoundly pessimistic, one can consider whether pessimism, as such, is always negative; Nietzsche in his preface to *The Birth of Tragedy* claims that the ancient Greeks required the "art-work of pessimism" in order to evolve into a higher consciousness:

> . . . Is there a pessimism of *strength*? An intellectual predilection for what is hard, awful, evil, problematical in existence, owing to well-being, to exuberant health, to *fullness* of existence?

—and the obvious *Yes* to these queries leads us into one of the great works on tragedy, which seeks to define it in terms of the issue Nietzsche would develop throughout his life, the relationship of the individual as Creator to the vast process of evolution in which he participates. Nietzsche's vision is the fundamentally religious position that one cannot be allowed an "easy" belief; like Job, great suffering must attend and strengthen faith. But Nietzsche's faith in a tragic joy, in an awakening of stopped-up Dionysian wonders by the sheer violence of external events, is not at all Shakespeare's—as Tolstoy believed, the natural religious temperament, the mystical as opposed to the institutionally religious, is somehow missing in Shakespeare; one finds nobility, stoicism, momentary alliances like that between Lear and Cordelia, in which human love is celebrated, but the Dionysian energies in themselves are felt as dangerous, chaotic, and never healthy.

When Matthew Arnold spoke of the assumption by poets of the religious and philosophical function,[4] he anticipated a coordination of moral and intellectual faculties that would allow one to distinguish between aesthetic values on one hand, and the "unconscious poetry" he saw in the religious temperament on the other; otherwise he would not have been as optimistic as he was. For, without the psychological experience of which the "religious" attitude is an intellectual result, the pessimism of certain great works of art is experienced apart from the ritualistic impulse that allowed them to be, originally—if ever—affirmative. And we come to accept as a universal statement about the condition of man what the artist knows to be, from the inside, hypothetical and sometimes playful variations on a theme.[5] Above, I grouped Desdemona and Cordelia along with Edmund, not meaning to eradicate the traditional divisions (at least in *Lear*) into "good" and "evil" camps, but to suggest that, for the artist, a more important consideration is whether or not he can locate any crevices, any openings, any fountains in his work, through which the life-force can move, regardless of moral distinctions. The Unconscious supposedly does not recognize socially accepted distinctions of good or evil, but craves only some form of organism-centered completion, the release and celebration of energy in some form—and, though the art-work is infinitely more complicated than the biological organism, the need to push forward, to violate the existing homeostatic condition, is just as natural, just as relentless. Allowing for the restrictions of the era, which are not always antagonistic to the individual, the art-work becomes the public vehicle for the artist's private vision; and the more melodramatic the better, since the form of dramatic conflict best parallels the conflict of the personality's various elements, conscious or unconscious contents that can never reach a stable equilibrium so long as life continues. (Questions of haphazard organization of scenes, unlikely disappearances and reappearances in Shakespeare's plays, as in contemporary films, are relevant only to the experience of these works on the printed page; as visual spectacles, which release emotions in a sequence of scenes, they need answer to the same logic as our dreams, which they very much resemble.)

Whether tragedy in its "highest" form is really affirmative, or only worked, historically, to frighten its viewers into an intellectual affirmation of the status quo, there is no doubt that individuals in our time experience it as pessimistic, regardless of what they have been taught. The naïve response is, after all, one's best expression of human instinct. One does not analyze a dream in order to know what sort of emotions to feel about it; one uses the emotion to seek out the meaning, inseparable from the experience itself. Thus, *Lear* is profoundly pessimistic for us in the twentieth century, and we cannot know or approximate its value to the past. Once we distinguish our intellectual expectation of emotion from our actual emotions, we are prepared to approach a work of art from our own point of view, and only by this method can we discover what might be timeless in it.

GODDESSES

That moments of transcendence must be followed, and dramatically, by catastrophic endings is part of the fabric of tragedy; one might speculate that an art-form that is in itself predetermined will most convincingly present a world-view that is predetermined—in contrast, for instance, with the greater freedom of the realistic forms of drama and fiction that have followed Shakespeare's time. Where formal freedom is enjoyed by the artist, freedom is more likely to be enjoyed by his characters, though the evolution of "freedom" in its various aspects is always related to the historical moment.

However, the incompatibility of the visionary and the tragic in *King Lear* is excessive even for tragedy, and a way of isolating and analyzing the terms of this incompatibility is by noting the work's presentation of women: goddesses, all, but of a totally unpredictable and possibly terrifying nature.

The world of *Lear* is one in which the particularized, personalized human being finds himself in some contention with his role—a representative of his species, his rank, his "place"—King, Father, Everyman, God-on-Earth; Daughter; Bastard; Loyal Servant; Madman; Traitor. The terms in which he dramatizes these roles soon become uncontrollable by him, though he imagines initially—as Lear certainly does—that he is in absolute control, and even the wise Cordelia miscalculates her power to absorb the violent emotions in her father which she has provoked; it is not so much raw aggression that leads to tragedy, but the loss of control that results from a simple refusal on the part of a "character" to conform to a "role." Hence, the youngest and fairest daughter of the king refuses to be the daughter of a king, but insists upon speaking as a woman who is Cordelia, and no other. In the acknowledgment of a separate, unique destiny, a personality possessed *not* by the sovereign but by the individual, there is a hint of the Void: formless horror, the music of the spheres violated, the unstoppable upheaval of raw nature. In this woman's insistence upon a moral intelligence not determined by her social role we have rebellion, the first and the most surprising of all. The others are for gain, for power, for exciting, new, lustful alliances, but

Cordelia's is without any ostensible purpose: she declares herself unwilling to lie, she declares *herself* as a self.

The "self" of Lear, however, is overwhelmed by the authority of the "King," in the grip of the most primitive of emotions, a human being dying inside an archetype. By the time of Lear's redemption, however, from this ignoble self, what is mortal in him has been lost to any role that might be accommodated in the structured world of man—of politics, of history. Shakespeare's cynicism is darker than one thinks, at least in *Lear*, for, though one may be broken upon the wheel of betrayal—the denial of Kingship by both a kingdom's subjects and by Nature itself—and "cut to the brains," the only knowledge he returns with is the knowledge that one *cannot* operate sanely in that place where "poor rogues/Talk of Court news." The necessary withdrawal of the enlightened man from politics, from the world as it exists in history, must have seemed to Shakespeare the only way in which a measure of transcendence, or true "selfness," could be retained. And yet—to surrender the world to those who demand it, precisely those who should not possess it! Part of the play's terrible pessimism is due to this assumption of a (saintly) passivity in the face of history, as if politics, the world, history, time, contaminated the morally virtuous: an assumption that is probably quite psychologically valid for most people, and yet presents, in art, an intolerable paradox.

However, having detached himself from the "role" he had been cast in, having fled into and through Nature itself, Lear satisfies our emotional demands for a dramatic rejection of the ego (by way of rejecting the superficial, time-determined roles of that ego), and his loving alliance with Cordelia suggests a wedding of sorts, an embrace of contrarieties: male and female, civilization and "great creating nature" rather than nature in its evil sense. A critical approach that examines the play as a coherent narrative, dealing with fully realized psychological events, arranged in a causal pattern, may be quite rewarding in that it satisfies our uneasy wishes that a work of art make sense on the most fundamental level, but it may be ultimately self-defeating; for one cannot disagree with Tolstoy, who was angered by the absence in Shakespeare's work of recognizable human beings, as well as the multiplicity of "unnatural" events—one may only disagree about whether these elements are always essential. It is impossible, now, knowing what we do about the effects of environment upon all human beings, including artists, to pretend that a work may not be valuable precisely in what it omits, what it rejects, what it demonstrates as unconscious assumptions unconsciously given voice in the externalization-process that is art.

One of Lear's more desperate passions is to know whether there is "any cause in nature that makes these hard hearts" (III, vi, 75). His fate is to learn that there is, there must be, since the hardness of hearts unites (in Shakespeare's imagination) man with nature, and nature must always be chaotic because it is not the Court, because it is not Art—it promises no immortality because it has no memory. The very form of the sonnet is outrageously and shamelessly egocentric, and allows the ego a pleasure that somehow activates guilt for its very celebration of form and language: hence the sonneteers acknowledge their constant terror of death, by guaranteeing themselves and their patrons the word "immortality," if not the condition. Confronted with the ungovernable processes of nature, many men—and not just the baffled, infuriated Lear—imagine that their "wits begin to turn." For nature when it is Nature, when it is experienced as outside the human ego, the human intellect, the human capacity for tyrannies of any kind—the most subtle, the most winning, the tyranny of language itself—is always the enemy, always fallen; and if animals are evoked they are not animals, but "beasts," and we experience the rage of authoritarian disappointment in terms of savage wolves, tigers, serpents, vultures, kites, adders and insects, rats, and "mad" and "biting" dogs.

Tragic enough, certainly, yet the ultimate tragedy is the experiencing as "enemy" the entire female sex, *even* one's dead and buried and presumably docile queen. The dilemma is that, for both Lear and Shakespeare, redemption must come only from the female, temporarily exiled in France, but required—and so pragmatically, as well as instinctively—in order that some measure of salvation be assured. If there could be a force or a being somehow uncontaminated by nature, a creature immaculately conceived, perhaps, then Man might be saved; the old kingdom restored. But there is only one savior possible, Cordelia: that one daughter of Man who, in the anonymous gentleman's words, "redeems Nature from the general curse/which twain have brought her to" (IV, vi, 202). Yet Cordelia is a woman, and as a woman she is Nature; she will not die and so she must be murdered.

Shakespeare deliberately alters the ending of the Lear story, in order to defeat the very salvation his work, from the inside, requires; it is not necessary to assume, as some critics do, that Shakespeare was projecting his own revulsion for women into the play, but it seems necessary to assume that whoever came to embody Nature, whoever spoke and acted freely, spontaneously, *naturally*, and rejected the archetypal role in order to affirm individuality, must be murdered—her magical powers, undeniably wonderful, stem from Nature and are therefore dangerous. Harbage notes that Shakespeare alone "and in defiance of precedent conducted Lear to ultimate misery"; pre-Shakespearean forms of the story ended happily.[6] One feels that he acted in *defiance* not only of precedent but of the unconscious folk-wish the play surely dramatizes, that the mortal ego be reunited with its soul, its own capacity for divinity, felt as such an irresistible psychological necessity that, as everyone knows, and imagines to be absurd, Nahum Tate rewrote the conclusion in 1680, in the order that Cordelia and Edgar might marry: if not the old man, then at least let Edgar have her!—the folk-impulse gratified, and yet curiously unworkable. The play is so baffling, so unconvincing, and yet so unforgettable, precisely because there is no conclusion possible at all, given the premises of the problem Shakespeare set himself—that fallen Nature somehow engenders a being not corrupt and not fallen, a savior. It was an impossible task. And, while the play is remarkable, even for a Shakespearean play, in its disregard for verisimilitude, the offstage event in which Cordelia is "killed" seems to me unimaginable from any angle. One cannot visualize that scene, not even with the greatest good will, for it requires us to believe that a soldier might enter Lear's and Cordelia's cell, noticing neither Lear nor Lear's agitation at his daughter's hanging—that Lear wait as the soldier hangs his daughter, and then that he spring to life, and murder the soldier. It is so preposterous a scene, even in an allegorical work, that had Shakespeare wanted to bring it into the dramatic action he could never have made it work—not delicacy but good sense required that it be kept offstage, like the Greek catastrophes it seems to parallel. (It is unfair, of course, to analyze a poetic work in terms of naturalism—but perhaps justified in this unusual case, since Shakespeare himself invites us to question that ending, by daring to force it out of its natural curve toward redemption.)[7] What is "cheerless, dark and deadly" is the conception of Nature as antithetical to Art or Artifice, and this curse determines the tragedy, quite apart from characters and their motivations and actions. Great art usually allows the instinctive life its articulation on a high, aesthetically satisfying plane: in *Lear* the very life-force itself is denied, and it is impossible to see the work as "religious" in any way.

Yet Arthur Sewell, along with other scholars and critics, would defend the play against charges of nihilism; Sewell even goes so far as to ask, "Does not the play look forward to Dostoyevsky, rather than back to Seneca?"[8] How peculiar, to have read Dostoyevsky in such a way that the possible death of Sonia or Alyosha could have been entertained—to have misread Dostoyevsky as a tragedian, rather than a mystic, whose vision of mankind is comparable to Dante's and whose "comic" side could accommodate a saint who disappoints his adolescent worshippers by beginning to smell quickly after his death—yet is no less a saint, for embodying nature's caprices. What Dostoyevsky and Shakespeare certainly have in common, along with their genius, and their fantastic imaginations, is the belief that suffering democratizes and allows growth and the awakening of wisdom; but this is not a "tragic" view necessarily. Folk-art teaches us the same thing.

Yet there is no single man, no single "Shakespeare"; Anthony Burgess's novel, *Nothing Like the Sun,* for all its gorgeous language, bitterly disappoints us in its portrayal of only the Shakespeare of the darkest plays, ignoring the Shakespeare of *The Tempest.* And in this we see how difficult, how very nearly impossible it is, for the serious artist to deal with the religious, affirmative spirit, or even with the phenomenon of a changing self, a self in flux. The critic must limit himself, in all honesty, to speaking only of the author of the work before him. Therefore, though I use the name "Shakespeare" I am really referring only to the author of *Lear,* a temporary personality, yet one in which many of the inclinations revealed in other works (in *Hamlet* and the sonnets, for instance) are given specific, savage voice: the wholesale denunciation and destruction of the female element, though this action will result in the thwarting of the tragic element itself, and the play as a whole will impress us as the aesthetic equivalent of a suicide. (*Re-enter Lear, with*

Cordelia dead in his arms: mortal man, his soul dead inside him.)

Because Shakespeare was a dramatist, it was natural that he perceive his characters more from the outside than the inside, as "actors" in a total spectacle, and that he force their individual personalities into roles, especially when he dealt with history. The more *individual* a character is, like Hotspur, Falstaff, Edmund, Mercutio, and the irresponsible Prince Hal before he becomes the responsible and priggish Henry V, the more it is necessary to subdue him, to annihilate him or transform him so that, at the play's conclusion, the audience is left with a single impression. One can interpret this from a pragmatic point of view—all professional dramatists are wonderfully pragmatic—or, as G. Wilson Knight does, more sympathetically, as Shakespeare's attempt to create a "poetic wholeness" that allows in a work like *Lear* "the most fearless artistic facing of the ultimate cruelty of things in our literature."[9] For Knight, *Lear* is a great work in that it confronts the very *absence* of tragic purpose, and that it gives us a tragic purification of the "esentially untragic."

Whether Shakespeare's *Lear* is an intensely private vision of evil, or whether the joining-together of the two stories and the alteration of the ending is a dramatist's private attempt to outdo earlier versions, or whether *both* possibilities are operating here, one cannot tell: we are left, however, with no single personality in the play that is not firmly trapped in "Nature," since only Edgar and Albany survive, and the single means by which Nature was to have been redeemed is dead. All is subdued to this conclusion, which bears little resemblance to the cathartic and rejuvenating conclusions of more conventional tragedy. Edmund may have contemptuously rejected the planetary influences (as the doomed Hotspur also rejects them), but Shakespeare dare not reject them as a dramatist, for to do so would be to strengthen his rebels' sense of freedom. When Shakespeare himself is freer, in terms of sympathizing with both sides of a conflict (as in *Antony and Cleopatra,* and *Troilus and Cressida),* it is important to note that he tends toward cynicism, rather than the more truly tragic realization of, for instance, Aeschylus in *Libation Bearers*—that "Right clashes with right." In *Lear* he suggests a tragically false dualism: Edmund's "Goddess," raw nature as interpreted by a bastard son of instinct, by which is meant sheer anti-social egotism, and, by contrast, the asexual "Goddess" it is Cordelia's fate to give life to, and to die in. She is also her own father's "soul in bliss," the perfect savior and the perfect victim. As Lear's unrepressed "inner voice" she speaks defiantly before the Court—the world—like another Eve involving us in another Fall, an unfortunate dividing of the kingdom into two and not into the mystical, indissoluble three. The "promised end" is the Apocalypse, in one sense; in another, the inevitable horror that follows when Nature (or woman) is given the freedom to act spontaneously, to upset ritual, rising in rebellion against masculine authority. All the "goddesses"—the "good" Cordelia, the "evil" Goneril and Regan—must die, the kingdom must be totally purged of the female, not in order that mere evil be eradicated, but that the life-force itself be denied. *Lear* generates excitement through its dramatization, in fantasy, of the suicidal wishes that lie behind all political and moral repression.

A KINGDOM WITHOUT A QUEEN

The disgust expressed in the play toward women is more strident and articulate, and far less reasonable, than the disgust expressed in *Othello* and *Hamlet* and certain of the sonnets. In other works, *Antony and Cleopatra,* and the comedies, women are allowed a certain measure of equality with men, but only through having lost or rejected their femininity; though Cleopatra is alluring, a temptress, we are shown the ways by which she deliberately calculates her triumph over Antony's defenses, and she emerges as more of a comrade, an "equal," not an intensely feminine and therefore magical (the interpretation is Iago's) woman like Desdemona, whose very innocence is fatal. In *Othello* and *Hamlet* and in the sonnet sequence, sexual loathing is in response to real or imagined infidelities on the part of beloved women; in *Lear*, however, sexual loathing is only a part of the general fear and loathing of Nature itself, most obviously represented by women. Cordelia is virginal and all but sexless, yet she is no less a woman, "a wretch whom Nature is ashamed/Almost to acknowledge hers" (I, i, 215).

Lear goes on to rail against Goneril and Regan as if their attitude toward him, in subsequent scenes, sprang from something inherently feminine in their nature, even something erotic; but in fact both daughters are behaving toward the old King, at this point in the play, like rebellious sons who are testing their father's authority. There is nothing feminine about them at all, and in the original Lear story in the *Arcadia* it was really Lear's sons-in-law who rebelled against him in order to get his kingdom, not his daughters. But Shakespeare deliberately goes against his source and makes both daughters enemies, and Albany a sympathetic character. In order to give a poetic wholeness to the anti-feminine brutality of the play, it was necessary that Shakespeare do this; in a causal sequence, Cordelia initiates the tragic action, her sisters continue it, her sisters die, but their evil continues so that Cordelia herself is executed, as a consequence of feminine rebellion of one kind or another. Edmund, of course, behaves in an evil way toward his father, but we are told that he is a bastard who has sprung from some "dark and vicious place" (that is, an unmarried woman's womb) and that Gloucester's succumbing to sexual instinct, so many years before, has now cost him his eyes.

Intolerable as female evil is to men, yet for some reason it cannot be easily annihilated, as Albany laments:

> See thyself, devil!
> Proper deformity seems not in the fiend
> So horrid as in woman.
>
> . . .
>
> Thou changèd and self-covering thing, for shame,
> Bemonster not thy feature. Were't my fitness
> To let these hands obey my blood,
> They are apt enough to dislocate and tear
> Thy flesh and bones. Howe'er thou art a fiend,
> A woman's shape doth shield thee.
>
> [ALBANY to GONERIL, IV, ii, 59ff]

So, while women like Goneril and Regan do not hesitate to obey the promptings of their "blood," like the bastard Edmund, a truly noble man like Albany does resist—for though such evil is obvious, it is shielded by "a woman's shape."

In purely metaphorical terms, Cordelia's natural mate would be Edmund: both are those dangerously spontaneous children, those outcasts, through whom the life-force leaps so explosively. But in terms of the plot Edmund is the mate both sisters desire, implausible though it is that such fiendish creatures could succumb to genuine love—for love it is, and not simply lust, since no man or woman ever chose to die for lust:

> GON. (*Aside*) I had rather lose the battle than that sister
> Should loosen him and me.
>
> [v, i, 18–19]

No attempt is made on Shakespeare's part to account for the sentimental rivalry over Edmund that would lead the vicious sisters to such extreme statements, and to death, for though Cordelia is granted the transcendence of the flesh that makes her into a "soul in bliss," her sisters are seen in these famous terms:

> Down from the waist they are Centaurs,
> Though women all above.
> But to the girdle do the gods inherit,
> Beneath is all the fiends'.
> There's Hell, there's darkness, there's the sulphurous pit,
> Burning, scalding, stench, consumption, fie, fie!
>
> [IV, vi, 126ff]

It is not dramatically clear why the sisters' cruelty to their father should be related to sexual desire, or why Lear should speak of "divorcing the tomb" of his dead wife, unless madness may be used to account for all his excesses. Yet he is not "mad" in the first act of the play, in which he threatens Goneril with the "kindness" of her sister:

> I have another daughter
> Who I am sure is kind and comfortable.
> When she shall hear this of thee, with her nails
> She'll flay thy wolvish visage. Thou shalt find
> That I'll resume the shape which thou dost think
> I have cast off forever.
>
> [I, iv, 327ff]

The bestiality of women, then, is not an absolute; when it is in the service of the King it is "kind and comfortable." What is *absolute* is the King's authority—even when he is raging, when he is mad—so that Gloucester quite naturally asks if he may kiss Lear's hand, after the impassioned curse quoted above, which compares women to Centaurs, and Kent's buffoonery before Gloucester's castle is honorable. It is a world in which the masculine archetype can do things wrongly, and yet never embody wrong, and in which the highest embodiment of the feminine, Cordelia, is represented as totally selfless, the perfect sacrifice.

One of the strangest interpretations of Cordelia's role is Freud's, in an early essay (1913) called "The Theme of the Three Caskets." Freud argues that Cordelia, as the third daughter, is Death itself, and that the "silent goddess" who destroys Lear is the last of the three forms his relations with women must take. Since nearly everything in Freud's cosmology is related back to the Oedipal complex, it is not surprising that Lear, an elderly patriarch who manages to attain a true transcendence of his *personal* miseries, should nevertheless be seen in these reductive terms: ". . . it is in vain that the old man yearns after the love of woman as once he had it from his mother; the third of the Fates alone . . . will take him into her arms."[10] Since Freud tended to equate the "feminine" with the "Unconscious," and both with those contents that threaten civilization, and the masculine ego, with dissolution, he is led to the extreme of reversing the play's general insistence upon Cordelia as life-bearing and spiritual, rather than a deathly embodiment of the Earth Goddess, and his interpretation cannot possibly account for the play's conclusion, in which the old man appears with Cordelia dead, in his arms. Cordelia as a form of Death cannot be supported by any evidence within the play, in terms of poetic imagery, for she is not only dissociated from raw, unspiritualized passion, but Lear is led to speak of her, at the play's conclusion, as dead as *earth* itself—so that she seems to us as far removed from the Magna Mater, the Terrible Mother, as it is possible for a female character to be. It would not be ironic that she is dead as earth itself, if "earth" had been, in any way, a suitable metaphor for her. What is curious is that Freud does not remark upon the imbalance of the kingdom—the one-sidedness of a kingdom ruled only by a king. A psychology that has as its model a balance of male-female, or "masculine-feminine" characteristics, might have speculated that "tragedy" issued from such one-sided development, both in the individual and in culture. Freud's psychology, of course, does not have this kind of balance as a model.

King Lear strikes us, at the same time, as an experimental work—one that poses and tests a vision of life necessarily related to the social and political milieu of the times (in which intrigue, hypocrisy, scandal, and murder were commonplace), but timeless in its anguished tension between what is "natural" and what is "unnatural" in human experience. How, given the savage terms of the play's universe, can man be redeemed from a partial, one-sided, blind fate? —pulled in one direction by the archetypal role he must play, and in another by a human, emotional, instinctive need that cannot be suppressed, or expressed, without violent consequences? Scholars suggest that the play was written sometime before December 26, 1606, but probably after the death of Elizabeth in 1603—after the death of a queen; and the work is characterized by a nightmarish sense of peril, of impending apocalypse that has nothing to do with the masculine hierarchical world, but stems directly from nature itself:

> GLOU. These late eclipses in the sun and moon portend no good to us: though the wisdom of nature can reason it thus and thus, yet nature finds itself scourged by the sequent effects: love cools, friendship falls off, brothers divide: in cities, mutinies; in countries, discord; in palaces, treason; and the bond cracked 'twist son and father. . . . We have seen the best of our time: machinations, hollowness, treachery, and all ruinous disorders, follow us disquietly to our graves.
>
> [I, ii, 111ff]

True, no doubt: as it seems generally true today, and true for all times, since the Apocalypse as a form of collective ego-despair and ego-love is always imminent, and always expressed by an era's imaginative artists in such terms. Yet for some reason the feminine forces are—if not in actual league with—not so vulnerable to the sequent effects. The play issues a stern, puritanical warning to all men: if one strays outside the harmonious structure as it is realized by men, if one descends to that "dark and vicious place" where the bastard Edmund is conceived, civilization itself will be destroyed.[11] The wheel will come full circle.

GRACE

Lear is experimental as well in its dramatizing of the soul's yearning for infinity, the desire of man to reach out to a higher form of himself, if not actually to "God" (Shakespeare's atheism seems unarguable). In purely psychological terms, Lear is the incomplete personality, the immature adult, forced by suffering to undergo a transformation that takes him far beyond himself. If hubris necessarily invites the death-blow of nemesis, the neurotic or unfulfilled personality *necessarily* indicates a higher self, the potentiality for fulfillment on a higher level that is totally lacking in contented, "normal" human beings, who have reached the end of their development. Clinical psychology and imaginative literature may or may not support a theory of the neuroses as unfulfilled contents of the self that are immensely valuable, and that are in some way related to unfulfilled elements in culture itself, but the aesthetic structure of a dramatic work is built upon the presupposition of change of some kind, in time; an incomplete condition is allowed its completion. In the melodramatic tragedy Shakespeare wrote, the latent villainy and the latent heroism of such a man as Macbeth are allowed their development, and the "man" who embodies them—the character who is called Macbeth—must be seen as little more than the vehicle, the metaphor, for that development. One is not given a character, Macbeth, whose psychological state leads him to certain acts of villainy, and ultimately to a kind of transcendental courage, but rather the illustrative acts themselves, flowering out of circumstances, to some extent "fated" by nature. G. Wilson Knight is surely correct when he stresses, in Shakespeare's tragedies, the significance of the pattern rather than the particles that make it up.

Lear demonstrates more powerfully than Shakespeare's other works the value of experience, even if that experience is suffering and death itself. In resisting and banishing the "Other," that part of the soul that is highest in man, Lear exaggerates man's natural tendencies to resist his own fulfillment, just as this tragic work exaggerates the literal dangers of such resistance: "I fear I am not in my perfect mind," Lear says, after he has been broken out of his "perfect" egotism, and succumbed to temporary madness. In order to complete his soul and be redeemed (in psychological terms: to activate his fullest identity) the hero must unite with the element that seems to oppose him. Because King Lear rules a world by himself, without a queen, his inclination toward the most dangerous of all masculine traits—tyranny—cannot be checked, except by the rebellion of a spontaneous intuition within the soul, but out of reach of the conscious mind. Hence, Cordelia, the youngest and fairest of the King's daughters, a part of his flesh itself, must oppose him. She is instinct's unsuppressable truth, required by Lear's one-sided soul; yet it is a supra-individual predicament, a one-sidedness that is symptomatic of Lear's culture itself, and not so readily cured.[12]

The vision Shakespeare might have been attempting in *King Lear* is the mystic's synthesis of self and "Other," time and eternity, the finite and the infinite, poetically symbolized by a union of male and female elements. Act IV shows us Lear asleep in the French camp, with "soft music playing"; when he is wakened by Cordelia he believes, at first, that he is dead, in hell, and that his daughter is a spirit:

> You do me wrong to take me out o' the grave.
> Thou art a soul in bliss, but I am bound
> Upon a wheel of fire that mine own tears
> Do scald like molten lead.
>
> [IV, vii, 44]

She tells him that he is "in his own kingdom"; the great rage of his former personality is now "killed in him." Cordelia functions as the embodiment of grace, that which is unearned, the redemption of the personality from the inside, out of the control of the conscious will. "Grace" is the usual religious term for this miraculous self-healing, but all of the healing sciences—medicine, psychology—are based upon the ability of the organism to heal itself, with or without the active interference of the will.

From this point onward Lear demonstrates a wholeness of personality that takes him beyond the nobility of soul possessed by any tragic hero in Shakespeare. He does not lust for revenge, but is prepared to "wear out,/In a walled prison, packs and sects of great ones/That ebb and flow by the moon" (v, iii, 17–19); he speaks of himself and his daughter as "sacrifices." Not until Cordelia is hanged does he commit any act of violence himself. When Lear carries Cordelia onstage, dead, Kent asks "Is this the promised end?"—that is, is *this* the end of the world, the Apocalypse itself?—and we feel that the "promised" completion in terms of the hoped-for rejuvenation of Nature has been totally thwarted, while the play's deeper movement, toward an eradication of all transcendental awareness that is predicated upon the feminine, has been brought to absolute completion. The Apocalypse serves man's purposes, for it brings together "Heaven" and "earth" but excludes the kind of raw, sensuous nature that Edmund worships. This "religious"—one might almost say *Protestant*—Apocalypse is not a mystical union of all of the universe, experienced as divine once history is suspended, but rather an expression of political rage, as in Young Clifford's words upon seeing the body of his dead father, in II *Henry VI:*

> O, let the vile world end,
> And the premised flames of the last day
> Knit earth and Heaven together!
> Now let the general trumpet blow his blast,
> Particularities and petty sounds
> To cease!
>
> [v, ii, 40ff]

"Ripeness is all": a statement of the body's limitations, and the need of the spirit to adjust itself, stoically, to such limitations. There is no visionary release from the body, or from history, and the play's ostensible hero—who will inherit the kingdom—seems to be saying, in these lines, that the vicious gouging-out of his father's eyes was somehow deserved:

> My name is Edgar, and thy father's son.
> The gods are just, and of our pleasant vices
> Make instruments to plague us.
> The dark and vicious place where thee he got
> Cost him his eyes.
>
> [v, iii, 169ff]

A puritanism that is so uncompromising draws the ideal into flesh only at the terrible risk of having to murder the ideal, because it *is* flesh: Cordelia, like Christ, is an inevitable victim. But it is unlikely that Shakespeare would say, as Milton did, that the Fall of Man might be justified—might even have been a *good*—since it brings the redemption, the divine into flesh. The Fall is not an event in Lear's world so much as a norm; one does not want to survive, given these conditions—Kent speaks of a "journey" he must take soon, indicating that, like Lear, he will not long outlive these images of revolt and chaos. To remain alive and rule the kingdom, as Edgar will, is a duty, an obligation only. The world has been emptied of all vitality, that of the soul's spontaneous rebellion against the ego, as well as that of bastardy and excess. Though Cordelia is murdered, one feels that the value she represents should not have been murdered; yet Edgar will rule the kingdom as Lear did, without a feminine counterpart.

Because the Lear stories concentrate upon the masculine predicament of kingship and fatherhood, and the dangers in relinquishing both forms of authority, it is dramatically necessary that the queen be already dead. Symbolically, however, it is the psychological value of the queen—the feminine—that is dead, absent, so that below the level of consciousness Shakespeare might have been led to attribute to that very absence a power for harm, dissolution, and terror: much as repressive and ego-fixated cultures tend to attribute to the suppressed elements (normal instinctive urges) an uncanny power. Within the individual, the melodrama is a familiar one, raised to tragedy when the instincts are so violently suppressed in the name of "rationality" that destruction results—aggression turned outward upon a usually innocent object of one's projected emotions, or aggression turned inward in the form of madness or suicide. In Erich Neumann's monumental *The Origins and History of Consciousness* the projection of "transpersonal" contents upon individual persons is discussed at great length, as well as the dangers to sanity that result from a helpless confusion of one's own person with the archetype one partly embodies. The patriarch's unspoken imperative, *Away from the unconscious, away from the mother,* is dangerous precisely because it is unspoken, unarticulated, kept below the threshold of consciousness itself. But, because the "unconscious" is so feared, the ego begins to project these fears upon the outside world, and so we have the common phenomenon of paranoia, which rages in those individuals who attempt to direct their lives *away* from the unconscious and in line with an idealized moral code. One of the extraordinary things about life—which Shakespeare's tragedies reflect so powerfully—is that while men of good will and intelligence can recognize the unconscious elements determining another's paranoia, they are invariably blind to their own projections; and, indeed, there is no way to determine what is real and what is simply projected, except insofar as one begins to experience intense emotions that are out of proportion to what other people are experiencing, given the same objective stimuli. The psychology of the puritan, the zealously moral man who overreacts to sin, and who is fascinated with sin, is only available to analytical study when his culture has developed away from him, so that he is italicized against it: so Shakespeare gives us that paradoxical but wise "dark comedy," *Measure for Measure,* in which repression itself generates the drama, but, in *King Lear,* it seems to me that Shakespeare was too involved in Lear's sexual paranoia to clearly delineate the psychopathology that has gripped the king. Very exciting it is, extremely convincing—Lear's dread of the daughter who will speak her mind, the chaos of nature that will not be governed, the female impulses that leap, uncontrolled, to the most forbidden of all objects, the illegitimate son; and it is exciting and convincing because Shakespeare feels Lear's passion from the inside.

When the feminine or maternal is not objectified, it begins to take on too powerful an essence. It "haunts" the conscious mind. Denied finite objectivity, the feminine is inflated out of all proportion to any individual's ability to contain it, just as any unconsolidated, unvoiced yearning becomes inflated and deadly, threatening to crowd consciousness out altogether. There is no clear dividing line between the harmless eccentricity that is one's "humour" and the obsession that ultimately drives one to madness—and the sense of bewilderment and gradual distaste we feel in reading such comically obsessive writers as Swift and Louis-Ferdinand Céline (both of whom seemed to despise quite ordinary natural functions) grows out of our not knowing, as readers, how serious the obsessions are. Dealing with them as "art," we are inclined to experience them with a certain detachment, and to imagine that the writers themselves felt this detachment—until we learn more about them through letters or journals. It is rare that an obsessive writer like Dostoyevsky (who hated Jews, Roman Catholics, and various "foreign elements") can produce works of art that avoid this violent identification of author and subject, and transcend limitations of the personal ego.

Ironically, Cordelia functions as that archetype of the soul, the sister or "anima," that is *not* maternal and that—in such forms as Athena and the Virgin Mary—represents a triumph over the Terrible Mother, the formless and all-devouring force of the unconscious that threatens dissolution; yet Lear (and Shakespeare, perhaps) responds to her initially as though she were an enemy. When she is banished, all of nature becomes suspect, and her two sisters—far closer to the "unconscious" instincts than Cordelia herself—rapidly degenerate. The primordial form of all godliness is the Magna Mater or the Terrible Mother who, like the Hindu goddess Kali, gives birth and devours without regard to individual achievements, personalities, gradations of consciousness: in short, the nightmare that threatens civilization itself. The "anima" figure, however, is intimately connected to the male, and is a helper of the male: so Athena springs full-grown from the head of Zeus, and does not require a woman in order to be born. Lear's three daughters have no mother, in a sense, but are *his*. Yet, because the very differing functions of the "anima" and the "Magna Mater" are confused, because all of the feminine contents have been imagined as evil, Cordelia is identified with the very force she should be defeating. In Neumann's words, the "activity of the masculine consciousness is heroic" insofar as it voluntarily takes on the struggle to raise itself out of ignorance,[13] but it is doomed to tragedy when the struggle is involuntary, when paranoia blinds a man like Lear and causes him to imagine enemies in those who love him best.

And so the value Cordelia represents does die with her. Though one may argue about whether the play's conclusion is "uplifting" or "depressing," it seems incontestable that the drama's few survivors experience it as an "image" of the horror of the Apocalypse—that is, an anticipation of the end of the world. We are left with no more than a minimal stoicism (though Kent does not intend to live) and an acquiescence to the "gods" as they punish "pleasant vices" with wholesale devastation that wipes out the innocent along with the guilty. For what purpose?—to turn the wheel full circle, it would seem, back to the primary zero, the *nothing* that is an underlying horror or promise throughout. As the Fool tells Lear in the first act: ". . . thou art an O without a figure: I am better than thou art now; I am a fool, thou art nothing" (I, iv, 211ff).

Nothing will come of nothing: a self-determining prophecy.

Notes

1. "I can declare anyone enemy or friend at will"—Hitler.

2. Hardin Craig considers the linking-together of the Leir and Plangus (Gloucester) stories as a "stroke of genius," in his edition of the *Works* (New York, 1951), p. 981, while G. B. Harrison speaks of Shakespeare's having "transmuted an old tale in which evil is punished and good restored into a tremendous and pessimistic drama," in which the Gloucester story underscores the tragic irony of the Lear story (in Harrison's 1952 edition of the *Works*, p. 1137), though the final

product remains difficult, perhaps a kind of poetic experimentation with imagery. However, Russell Fraser's commentary in his *Essential Shakespeare* (New York, 1972) suggests that the Shakespearean version of the stories leaves us with a sense that both have somehow been violated, and that "abnormal behavior is the norm" (380).

3. In "Shakespeare and 'The Revolution of the Times,'" *Triquarterly Special Issue: Literature in Revolution*, Winter-Spring 1972, p. 244. In the universe of Shakespeare's time there could be no true freedom, since men's actions had planetary significance, and could unhinge the entire cosmos; to defy this cosmic order initiates "tragedy."

4. Matthew Arnold, "The Study of Poetry" (1880).

5. And sometimes the variations are severely conditioned by the class to which the artist belongs, or has rejected; by the immediate events in his private life which are transfigured into objective and frozen attitudes in his art; by his patrons' or his audience's demands. In an essay largely concerned with the powerlessness of art to alter political conditions, Louis Kampf questions the validity of what he calls the "humanist thesis," which allows artists to "generalize from their *personal* concerns to those of all people. If the artist feels tragic, the sense of tragedy becomes *the* human condition." "Understanding the Concrete Needs of the Historical Moment," *Arts in Society: The Humanist Alternative*, Spring-Summer 1973, p. 66.

6. Alfred Harbage, *The Complete Pelican Shakespeare* (New York, 1970), p. 1060. Harbage acknowledges that the play "ends as it began," and that its main gift for an audience is the gift of "feeling pity." When Harbage states that *Lear* is "religious, as all great tragedies are religious" and that the brutal killing of Cordelia is therefore a "sacrifice" and not a mere turn of the screw, one experiences that sense of vertigo, that bewilderment one was taught never to express as an undergraduate: to declare a tautology a tautology is to speak, like poor doomed Cordelia, against the Institution, and risk exile. One may learn very little about the "great works," but a great deal about one's own time, by studying critical responses to those works.

7. When the instinctive pattern is violated, everyone suffers; there is no possibility of an authentic "catharsis." As Erich Neumann in *The Origins and History of Consciousness* (Princeton, 1971), states:

> So long as a content is totally unconscious, it regulates the whole and its power is then at its greatest. But if the ego succeeds in wresting it from the unconscious and making it a

> conscious content, it is—mythologically speaking—overcome. As, however, this content still goes on using up libido, the ego must continue to work at it until it is fully incorporated and assimilated. Ego consciousness cannot therefore avoid further dealings with the "conquered" content and is likely to suffer. . . . the ascetic whose ego consciousness has triumphantly repulsed the instinctual components that threatened to master him experiences pleasure with his ego, but he "suffers" because the instinct he has denied is also a part of his total structure. [348]

8. Arthur Sewell, "Tragedy and the Kingdom of Ends," in *Shakespeare: Modern Essays in Criticism,* ed. Leonard Dean (New York, 1957), p. 331. Dostoyevsky fuses archetypal and individual features because he experienced the world in this way and because, to him, Sonia might be Sonia, a prostitute, and also St. Sophia; Alyhosha might be the youngest son of a depraved nobleman, and yet a form of Christ. Sewell fails to see that it is through an examination of the vital *differences* between Shakespeare and Dostoyevsky that one can approach some valuable understanding of both.

9. G. Wilson Knight, *The Wheel of Fire: Interpretations of Shakespearian Tragedy* (New York, 1968), p. 174. Knight's deep, thoughtful study, at times more a meditation than an analytical work, was originally published in 1930, reissued countless times, and remains one of the finest works of criticism on Shakespeare, or on any subject. In his prefatory note to the 1947 edition of *The Wheel of Fire,* Knight states clearly his belief in dramatic relationships that take precedence over individual "particles," the "poetic wholeness" he finds in Shakespeare having a similarity to the emphasis placed upon pattern rather than permanence, in modern physics. Poetry is

> . . . pre-eminently a blend of the dynamic and the static, of motion and form; and, at the limit, the perfectly integrated man, or superman, is to be conceived as a creature of superb balance, poise, and grace. Interpretation is, then, merely the free use of a faculty that responds with ease, and yet with full consciousness of the separate elements involved, to this space-time fusion, or relationship, this eternity, of art, in which every point on the sequence is impregnated by the whole. [viii]

10. Freud, in *Character and Culture,* ed. Philip Rieff (New York, 1963), p. 79.

11. If the life-force is always to be interpreted as a threat, and in political terms any enforced change in the status quo is "unnatural" and therefore "evil," what is suppressed (whether instincts or human beings like Edmund himself) will always overwhelm the status quo eventually. There can be no perfect order, no permanent authority. It was not the Apocalypse that was coming, but the Puritan Revolution; so Jan Kott's claim that Fate, in Shakespeare, is "represented by the class struggle" makes sense of a kind, though one might say, paradoxically, that the "class struggle" in certain works (*Othello* as well as *King Lear*) accounts for the tragedy itself. But Kott can reimagine *King Lear* only in terms of grotesque dark comedy, an *Endgame* of the Renaissance. See Jan Kott, *Shakespeare, Our Contemporary* (London, 1964).

12. John Danby speculates upon the probability of the "Shakespearean breakdown of confidence" being a reaction against Elizabeth's "overstimulation of the cult of Gloriana"—see "The Fool and Handy-Dandy," in *Shakespeare's Doctrine of Nature: A Study of King Lear* (London, 1949). In *The Story of the Night* (London, 1961), John Holloway discusses *King Lear* in terms of being, among other things, a "rehearsal of the end of the world."

13. "The deflation of the unconscious, its 'dethronement' by the patriarchal trend of conscious development, is closely connected with the deprecation of the female in the patriarchate. . . . The association of the unconscious with feminine symbolism is archetypal, and the maternal character of the unconscious is further intensified by the anima figure which, in the masculine psyche, stands for the soul. Consequently, the heroic-masculine trend of development is apt to confuse 'away from the unconscious' with 'away from the feminine' altogether." See Neumann, p. 340. It is interesting to note that when Shakespeare abandons the world of grim political reality, of "history," he can translate these various tensions into *The Tempest:* Cordelia becomes Miranda, Edmund, Caliban, Prospero, not only authoritarian ruler but poet, creator, as well—which is to say God, omnipotent and all-forgiving. The maternal elements are absorbed into the paternal; Prospero *is* both everything and nothing, like Shakespeare himself. But only in fantasy: only on that island. In England, in Lear's time-tormented world, masculine consciousness must triumph over all opposition, including life itself.

"The Immense Indifference of Things": Conrad's Nostromo

In our activity alone do we find the sustaining illusion of an independent existence as against the whole scheme of things of which we form a helpless part.
Nostromo

Action—the first thought, or perhaps the first impulse on earth! The barbed hook, baited with the illusion of progress, to bring out of the lightless void the shoals of unnumbered generations!
Victory

In Conrad's major works and especially in *Nostromo* (completed in 1904), the novel that is probably his masterpiece, one encounters mysterious and exasperating contradictions —or are they paradoxes?—that might help to account for the misery Conrad experienced in writing the novel. One of the most self-conscious of modern novelists, as rigorous and relentless in his own way as Flaubert, whom he admired, and as fundamentally pessimistic as Schopenhauer, whose life-denying and misogynous philosophy may well have influenced him, Conrad sets up again and again in his novels dialectical struggles—melodramas of "opposites"— that cannot be resolved except through the destruction of both, and the necessary deaths or defeats of his central characters. If authorial intrusion assures us repeatedly that our personalities, our identities, our very *lives* cannot be realized except through "activity" (see Martin Decoud's thoughts just before he commits suicide), the same authority will insist, will dramatize most sensationally and cruelly (especially in *Victory*) that the very reverse is true: Heyst's bitter, skeptical father, a Schopenhauerian "destroyer of systems," turns out to have been quite correct in diagnosing action as the "barbed hook" that leads inevitably to man's destruction.

Idealists like Jim and Kurtz (in his "European" phase) and Charles Gould are not only willing to gamble their lives for the sake of their ideals—which are, of course, "illusions" to Conrad; they are willing, quite unconsciously eager, to gamble the lives of others, of entire communities. They are suicidal: if heroic at all, their heroism consists of an unconscious, and in Jim rather adolescent, inflation of the ideal of simple, direct, aggressive, "masculine" activity. Conrad admires them, in a way—but he cannot take them seriously, nor does he wish to present them as tragic heroes. Marlow, Stein, Dr. Monygham, and even Nostromo—whose cynical intelligence is easy to overlook—speak more directly for Conrad, their profound skepticism a necessary critique and check upon the frenzied activities of the "idealists." But if the idealists are horribly limited in vision, mistaking the "bait" of melodrama for reality, and thereby drawing into destruction any number of other, less credulous people, it is certainly the case that the skeptics offer very little; the more convincing they are, the more their intellectual cynicism seems to be *the* ultimate comment upon human existence—for does not Conrad himself share Martin Decoud's attitude toward Costaguana's politics, and Nostromo's fate in the "desperate adventure" of the novel?—the more, necessarily, the illusory activities of life itself are condemned, mocked, and finally rejected. The idealist brings death to the human community; but so does the detached, enlightened skeptic.[1]

Similarly, the "barbed hook" of sexual attraction is a temptation only when disguised in a truly "feminine" female, like Lena of *Victory;* she who effaces herself completely, who is eager to sacrifice herself in an ecstasy of love for her man —Conrad's apparently quite serious idea of a "heroine"— thereby, ironically, destroying love itself. In *Nostromo,* the fairy-like Mrs. Gould, usually dressed in white, always patient, gracious, and only at the conclusion of the novel rather surprisingly cynical, is an ideally "feminine" creature, the necessary spouse, the wife without whom Charles Gould's masculine-aggressive idealism could not be realized: the author says dogmatically of her that "it must not be supposed that [her] mind was masculine" (73, Modern Library edition of *Nostromo*). A woman must not be womanly, or human-like; she *must* be feminine; otherwise she cannot be a proper mate for a man.[2] At the same time, her femininity, because it is so uncritical, so frankly brainless, will assure the continued delusions of her mate, and perhaps even bring about his death, as in *Victory,* or the loss of his soul, as in *Nostromo.* An intelligent man might well be advised to marry another man!—but of course this is not possible. It is not even impossible; it is unthinkable. When the sexes are very different, then, "opposites" in every way, catastrophe lies ahead. However, when the sexes are not clearly differentiated, when, for instance, the marvelous "Mr. Jones" of *Victory* is revealed as effeminate, with "beautifully pencilled" eyebrows and a virulent distaste for women, one can be sure that catastrophe lies ahead—for he refuses, categorically, to play the game, refuses to be charmed by any woman, even Lena, and though Conrad may well share his sympathies, he *is* a villain.[3] The opposition of men and women stimulates a tragic *maya* of attraction; but the rejection of one's sexual nature, or sexual role, also stimulates a tragic drama. The community suffers death in either case.

(The community will be, in fact, demolished if the individual is allowed to reject the conventional roles of masculinity and femininity: no further generations will be brought out of what Conrad calls the "lightless void." Always, in literature at least, disaster is assured by desexualization; Lady Macbeth prays to be "unsexed" and filled "from the crown to the toe top-full/Of direst cruelty," Lear's Goneril and Regan are really rivalrous sons, not daughters, and out of their rejection of the feminine springs the inevitable tragedy.)

One encounters, then, in Conrad, certain paradoxical dualities, blatantly self-contradictory beliefs somehow held in suspension—until the concluding pages of a representative work, when chaos rushes in, is in fact welcomed in, that the drama be ended if not resolved. An artist, of course, must have faith in the activity of his art; otherwise he could not create it. In this, Conrad is rather like Martin Decoud, inspired by love to patriotism of a sort, even to jingoistic newspaper work and the dangerous adventure in the lighter; he is like Nostromo as well, embarked upon the most "desperate" adventure of his life—the writing of *Nostromo*, which seems to have nearly destroyed him. At the same time, he is, of course, supremely detached; he is godly, omniscient, capable of telescoping into two or three paragraphs (in the remarkable Chapter 10 of Part III of *Nostromo*, especially) any number of points of view, and historical incidents, which allow us to see the pathetic, doomed folly that is the individual's role, however passionately he may immerse himself in it. As a novelist, Conrad is always superior to his ideas. His works are most living when "ideas" as such are set aside, in order that his characters may bravely play their scenes: Decoud talking himself into falling in love, Nostromo embarking upon his swim to the abandoned lighter's boat. But, inevitably, their energies are curtailed by the author's larger, more general, and very restrictive consciousness. "If one has character," says Nietzsche, "one has also one's typical experience that recurs again and again."

"CIVIL WAR": THE COMPOSITION OF *NOSTROMO*

Nostromo is the greatest of Conrad's novels, not only because it is the most ambitious, the most populated, and the most energetic of the longer works, but also because it represents a challenge only a courageous novelist would attempt. The creation of small, tidy, "perfect" works of art is by no means as tempting to the serious novelist as critics might like to think; even Henry James, capable of beautifully imagined, exquisite little stories, became impatient with the "thinness" of such achievements, and preferred, to conciseness, a more organic form in which "the fine, the large, the human, the natural, the fundamental, the passionate things" (see *Notebooks*) might be explored. James attempted nothing like *Nostromo*, of course; he was never so bravely reckless, perhaps never so passionately involved in his material as Conrad; at the same time, he would have been wise enough to see that *Nostromo: A Tale of the Seaboard* could be completed only with enormous labor, for it does not really delineate a *human* experience. It utilizes all the devices of the conventional novel—plot, character, "theme"—as well as certain impressionistic techniques Conrad had innovated (in *The Nigger of the "Narcissus"* and *Heart of Darkness,* especially); and it is as old-fashioned, as didactic, as relentlessly moralistic a novel as any Victorian work. At the same time there is something radically experimental about it. Proust, whom Conrad admired, might write of the passage of time and the evocation of time past, of personal history, but he did so safely within the confines of his narrator's consciousness; there is no "history" in Proust beyond the existential, the experienced. In Conrad, however, "history" is the only reality, and the merely human personalities who help to bring about the founding of the Occidental Republic are, upon the founding of the Republic, no longer very real. The subject of *Nostromo* is not Nostromo, not Charles Gould and his wife; it is certainly not Martin Decoud, who dies midway in the novel. It is time itself: the inexorable process of time, which cannot be stopped, and cannot be fixed into an aesthetic structure.

The material is, then, necessarily intractable. One cannot create a work of art upon the premise that all human activity —whether the founding of a republic or the creation of a novel—is ultimately defeated by the mere passage of time. Unless of course the art-work is totally experimental and shot through with self-lacerating ironies, like Mann's *Dr. Faustus*—or, like Joyce's *Finnegans Wake,* an attempt to obliterate time itself by the act of sharing in its passage and affirming its basically cyclical nature. But Conrad is not Mann, and he is certainly not James Joyce. He is far more modest in his claims for art; his pessimism is rooted firmly in the nineteenth century's rather premature banishment of the spirit from the measurable, demonstrable, "scientific" material universe.[4] (Hence Conrad's condemnation of all ideas as illusions, simply because the Kurtzes of his experience were so wildly and fatally deluded.) The spiritual cannot, in Conrad, redeem the material; spiritual interests *cannot* bring material interests into the truly human sphere. The exuberant faith expressed in the famous Preface to *The Nigger of the "Narcissus"* (1897) has been tempered by Conrad's experience of the increasingly disorganized drama that is contemporary history, in which a "mere novel" (see the Note to the First Edition of *Victory,* 1914) can have little redemptive power.

It may be, also, that great works of art are necessarily flawed or incomplete because they represent, to the artist, one of the central mysteries of his life—that which cannot be resolved, but which *must* be explored. To triumph over the mystery that is at the heart of one's own existence would be, for the artist, the acquisition of a rare treasure—not the "silver of the mine" in its crude, material form, but the treasure of the soul, the realized, integrated soul itself. The riddle of Conrad's art, alluded to again and again but never deliberately, clearly stated, might well be: *How is tragedy to be avoided, when one's living, breathing self, defined in and by action, is always inferior to one's ideal, Platonic essence?* (In D. H. Lawrence, by contrast, one's essence *is* physical experience; though humanity is crucified upon the cross of the antagonism between "mind-consciousness" and "blood-consciousness," this is not necessarily tragic; it is the life-force itself, and strife, tension, incompleteness, and even suffering are good because they awaken us to life.) Conrad's vision is tragic; but there is no glory in man's submission to mystery. Man does not count for very much, in Costaguana or on the high seas, in the "immense indifference of things"; in the chaos of time that defeats personal history, and makes of Decoud's and Nostromo's heroism only material for Captain Mitchell's tedious, self-serving anecdotes. And Captain Mitchell himself is greatly aged—no doubt a little senile—soon to die, and be replaced by someone else, to whom Decoud and Nostromo and the founding of the Occidental Republic will be of little consequence indeed.

Unlike Virginia Woolf, Conrad seems to have wanted it both ways—he wanted to write a novel about personalities (why else the title, *Nostromo*?) and a novel about ungovernable, unpredictable, shapeless events, in which the process of time itself is the sole reality. *Nostromo* is so baffling and fascinating a work, rewarding even after endless readings, precisely because it is so unfocused. Since there is no possibility of its being rounded off, made into an architecturally successful work, its innumerable dramatic scenes and the astounding "life" of certain of its characters are all the more memorable. What can poor Nostromo do, having lost his soul to the silver, except drift into a very nearly autonomous, sensationally melodramatic narrative—the affair of Giselle, Linda, and the lighthouse?—which has little to do, except thematically, with the novel itself. In a world in which brute, formless events take precedence over any human activity, the "heroes" of one epoch necessarily dwindle or degenerate or stumble into the mock-heroics of the next, like dethroned gods who turn up in a new mythology as demons or gargoyles. Traditional tragedy may appear brutal, as it clears the boards for the next event, wiping out most of the cast—guilty and innocent alike; but it is thoughtful, even rather romantic, for one cannot conceive of a more mature Desdemona, a less passionate Hamlet, a genuinely senile Lear, or a modest, diminished, unregal Cleopatra. Nostromo, the man of the people, never meets "his" fate at all, as Kurtz and even Martin Decoud do; he simply dies, accidentally, mistaken for another, lesser man.

Like the leading citizens of Sulaco, Conrad is, in the writing of *Nostromo,* accursed by the very riches that should redeem him. The novel is too big, too broad, seething with too much life. How to assimilate so many people, so many events? How to force a radically new vision of man's place in the universe into the confines of a traditional narrative? In *A Personal Record* Conrad speaks of his struggle: ". . . for twenty months, neglecting the common joys of life that fall to the lot of the humblest of this earth, I had, like the prophet of old, 'wrestled with the Lord' for my creation" (98–99); "Failure" and "Astonishing" are words that can be applied equally to the work—"take your choice," says Conrad. The work is, probably, a "failure" if one must think along such lines; but it is certainly an "astonishing" achievement.

Written at a time when Conrad was suffering from physical pain, and from the humiliation of poverty, *Nostromo,* in its complexity and in the sardonic wisdom it offers, must have had an extraordinarily deep, personal meaning to Conrad; it was certainly not written with much hope of commercial success. And the labor of composition was more than customarily difficult for Conrad:

> I sit down religiously every morning, I sit down for 8 hours every day—and the sitting down is all. In the course of that working day of 8 hours, I write 3 sentences which I erase . . . sometimes it takes all my resolution and power of self control to refrain from butting my head against the wall.
>
> [in a letter of 29 March 1898, in Jean-Aubry's *Joseph Conrad: Life and Letters* (London, 1927), I, 231]

In a letter of 1903, to R. B. Cunninghame Graham:

> I am dying over that cursed *Nostromo* thing. All my memories of Central America seem to slip away. I just had a glimpse 25 years ago,—a short glance. That is not enough *pour bâtir un roman dessus.* And yet one must live.

To H. G. Wells:

> I, my dear Wells, am absolutely out of my mind with the work and apprehension of my work. I go on as one would cycle over a precipice along a 14-inch plank. If I falter I am lost.

To Edward Garnett:

> *Nostromo* is finished; a fact upon which my friends may congratulate me as upon a recovery from a dangerous illness.

And to William Rothenstein:

> What [*Nostromo*] is like, I don't know. I don't suppose it'll damage me: but I know that it is open to much intelligent criticism. . . . Personally, I am not satisfied. It is something—but not *the* thing I tried for. There is no exultation, none of that temporary sense of achievement which is so soothing. Even the mere feeling of relief at having done with it is wanting. The strain has been too great, has lasted too long.

Conrad always speaks, sometimes rather dramatically, of the fact that writing is not easy for him, but his struggle with *Nostromo* seems to have been unusually painful. Completing the novel is as much of a chore, and evidently as thankless an adventure, as saving the silver of the mine is for Nostromo himself. Conrad complains of being reduced to near-imbecility; of feeling that his brain has turned to water. And what is the act of writing, except the "conversion of nervous force" into words? Wanting to create a great work of art, aspiring to greatness of any kind, no doubt cripples one's enjoyment of the artistic experience—for it is now a means to an end, rather than an end in itself; and Conrad was always very serious about his art. But the novel presented near-insurmountable difficulties for other reasons, some of them mentioned earlier (it is mistitled: *The Silver of the Mine* would have been more accurate, and surely more poetically suggestive; even Don José Avellanos's *Fifty Years of Misrule* is a better title, acknowledging the supremacy of time and history over human endeavor).[5] As intensely realized experience is recounted as anecdote, and the literal value of the silver itself questioned—Mrs. Gould tells the dying Nostromo that the silver does not matter, after all, it would have been better for everyone had it actually sunk ("Marvelous!" Nostromo replies, sarcastically. "—that one of you should hate the wealth that you know so well how to take from the hands of the poor . . . ," 624)—

the novel's subject becomes increasingly blurred, and it is more and more difficult for Conrad to force any structural unity upon it. His temperament seems to have leaned toward parable or allegory, in which a relatively limited cast of characters is under perfect authorial control (as in the inferior *Victory*, which is comprehensible *only* as allegory, and baffling otherwise); never does he seem to have allowed one of his characters any measure of freedom. The fact that Martin Decoud is the most lively, the most intelligent, and by far the most sympathetic character in *Nostromo* seems to have made his fate a foregone conclusion. Decoud is too intelligent to be controlled—therefore he must be eliminated. He is too much like Conrad himself, an alter ego to be feared; and perhaps Conrad's determination to punish Decoud was one of the chief reasons the composition of *Nostromo* was so painful—and its completion so unrewarding.

MARTIN DECOUD: LOVER, PATRIOT, SUICIDE

> *In the most sceptical heart there lurks at such moments, when the chances of existence are involved, a desire to leave a correct impression of the feelings, like a light by which the action may be seen when personality is gone. . . .*
>
> Nostromo

Nostromo springs to life when Martin Decoud appears. Formidable and impressive as the early pages are, crowded with geographical facts and historical detail, there is nevertheless a certain murkiness about so much richness; the novel desperately requires a voice, a human voice, to bring it into focus. Young Decoud, the "sane materialist," allows Conrad to make wittily explicit the numerous ironies of

Costaguana and its deadly *farce macabre*. Thirty years old, a former student of law, a would-be poet, Decoud has spent most of his life in Paris; he becomes involved in the current crisis at first because it "amuses" him to do so, then for more complex motives, one of which is his love for Antonia Avellanos. Despite his intelligent rejection of political idealism, or any sort of idealism, Decoud is, ironically, the "founder" of the Occidental Republic—the presumably civilized nation-state which is the setting of *Nostromo*'s present action. He never really becomes Antonia's lover; he never realizes how he and Nostromo have succeeded in their desperate adventure; he becomes, after his death, just another character in Captain Mitchell's rambling history. And even Conrad turns against him—is not content with merely killing him off, but must pass judgment on his death, in the most wickedly dogmatic terms.

Most critics accept Conrad's authorial comment on Decoud. If Conrad insists that the young man is an "egotist," and his death is somehow bound up with his "egotism," on what grounds dare one protest? He is not only an egotist but a "supreme egotist."[6] He shoots himself in the chest and falls into the Gulf not because he is desperately lonely and exhausted and suffering a quite natural psychological crisis, but because, as Conrad says so bluntly, he is a "victim of the disillusioned weariness which is the retribution meted out to intellectual audacity"; his fate is to be "swallowed up in the immense indifference of things" (560). It is said of him also: "The brilliant Costaguanero of the boulevards had died from solitude and want of faith in himself and others" (555). And, should we not quite grasp the degree of Decoud's imperfection, it is said as well: "He had recognized no other virtue than intelligence, and had erected passions into duties. Both his intelligence and his passion were swallowed up easily in this great unbroken solitude of waiting without faith" (557).

That a novelist should so humorlessly and willfully punish one of his creatures—and that creature already doomed!—might suggest a certain crude, punitive quality in the novelist, which art usually obscures; it might suggest, also, that such overwriting, such overdetermination spring from the writer's attempt to dissociate himself from that character, as if that character revealed truths about the novelist he could not quite accept. Martin Decoud's "egotism" is largely a matter of Conrad's rhetoric. Nostromo's marvelous vanity is dramatized, but Decoud's is not; and even if he were guilty of that elusive sin of "egotism," who in Costaguana is not? The selflessness of Mrs. Gould is shown to be sterile, a pointless sublimation of her energies; the idealism of old Giorgio Viola is not only anachronistic, but positively absurd in this part of the world; and Dr. Monygham's devotion to Mrs. Gould and the suppression of his own self are a result of his self-loathing. Presumably egoless are characters like Captain Mitchell, who are absorbed by the capitalist enterprises they represent, or fanatics like Father Corbelàn, allied with an ecclesiastical authority that would like to reclaim Costaguana for the Roman Catholic Church, the strife of politics being inconsequential, as are moral pretensions. The megalomania of a man like Charles Gould goes largely unnoticed because it is cultural rather than personal; rather like that other deluded idealist, Kurtz, Charles Gould states unblushingly that he *has* faith, and that his faith will be the salvation of the country. ("What is wanted here is law, good faith, order, security. . . . I pin my faith to material interests. Only let the material interests once get a firm footing, and they are bound to impose the conditions on which alone they can continue to exist. That's how your money-making is justified here. . . . A better justice will come afterwards" (92–93).) By contrast, the simple, unpretentious vanity of Nostromo seems almost heroic, and the shrewd skepticism of Martin Decoud the necessary voice of sanity. Decoud must die for reasons of the plot, probably, but why must he die ignobly?—why must he be so earnestly repudiated by his creator?

J. I. M. Stewart makes the important point that, for Conrad, the serio-comic miseries of Costaguana constitute a reimagining of the "emotional hinterland" of the Poland of Conrad's youth,[7] and Conrad tells his readers, in the Preface, of his still-abiding infatuation for "the beautiful Antonia"—she who is modeled after his memory of a first love, a young Polish patriot whom he had evidently not known very well. Conrad's language is sentimental, or icily ironic: he imagines Antonia in the great cathedral, "absorbed in filial devotion" before her father's monument, then casting a "tender, faithful glance" at the medallion-memorial of Decoud, before going out "serenely" into the sunshine of the Plaza. All genuine passion is dead, all life is dead—converted into monuments, medallions, statues, marble shafts topped with gilt. (Martin Decoud, isolated, dying, imagines his beloved Antonia as "gigantic and lovely like an allegorical statue, looking on with scornful eyes at his weakness" (556).) It was not the case, of course, that Conrad "lost" his Antonia, any more than that he had died in the quixotic service of Poland. If he is embodied in any aspect of the Decoud-Antonia relationship, it is only through desperate concern of Decoud that he convert passion into idealism—"passion" itself being inadequate as a motive, or forbidden. (Just as Heyst rescues Lena for altruistic reasons, as he had earlier rescued poor Morrison; and both "rescues" cause trouble.) In Conrad, very little is done directly, for motives that make biological or economic sense, that is, Conrad insists that the actions he portrays—even the senseless vengeful action of a Gentleman Brown, for instance, at the conclusion of *Lord Jim*—be related to attitudes of will or self-righteousness or "egotism." If there is guilt to be expiated, if Conrad is troubled over his own talent, brilliance, and undeniable skepticism, as well as his ambiguous feelings about Poland, one of the techniques whereby release from this guilt might be gained—temporarily—is that of imaginatively projecting into all his intelligent characters the need, very nearly the lust, to be thwarted, destroyed, or to actually commit suicide; the more a character approaches Conrad's own position, the more cruelly must he be repudiated.

The "egotism" of Martin Decoud is nothing more than his intelligence. That he is witty, sarcastic, that he sees through the pomposity of government "ideals," that he is, even, skilled in the manipulation of words, sets him apart from his countrymen, who are "mute" and "suffering"—and this distinction between types of personality makes Conrad uneasy, for Conrad is not mute, not inarticulate, not one of the "children of the sea" whom he has attempted to eulogize elsewhere, most didactically in *The Nigger of the "Narcissus."*[8] Everything Martin Decoud chooses to comment upon strikes us as perceptive; he is, in fact, more reliable and more human than Marlow himself (whose rhetorical excesses he avoids), and surely Conrad means to speak through Decoud in such passages:

> Imagine an atmosphere of opera-bouffe in which all the comic business of stage statesmen, brigands, etc. etc., all their farcical stealing, intriguing, and stabbing is done in dead earnest. It is screamingly funny; the blood flows all the time, and the actors believe themselves to be influencing the fate of the universe. Of course, government in general, any government anywhere, is a thing of exquisite comicality to a discerning mind; but really we Spanish-Americans do overstep the bounds. No man of ordinary intelligence can take part in the intrigues of *une farce macabre.* [168]

There is a curse of futility upon our character: Don Quixote and Sancho Panza, chivalry and materialism, high-

> sounding sentiments and a supine morality, violent efforts for an idea and a sullen acquiescence in every form of corruption. [189]

That so undeceived a man should, nevertheless, be drawn to an involvement with his country, should be willing to risk his life for it (or for "love") is certainly to his credit; his fall into fateful, necessary action, in contrast to his former detachment, parallels Heyst's, but Martin Decoud is far more complex than Heyst—had Conrad chosen to develop him, he would have been one of the most subtle personalities in modern fiction, a genuinely tragic figure. The difficulty with *Nostromo,* however, is its diffusion of energies; it is a magnificent attempt to create a work of tragic dimension, but without the necessary creation of significantly tragic personalities. "The Silver of the Mine" is the real subject of the narrative—but the "silver of the mine" is a vapor, has no identity, no symbolic or literal value whatsoever, apart from the activities of the human consciousness. Charles Gould, who disobeys his father's wishes and reopens the San Tomé mine, and who thereby disturbs the "snakes" slumbering in paradise, never comes to the point of recognizing, as Decoud does, the irony of his position; if anything, he becomes more intrigued by it, less human. It is Decoud who speaks for Conrad, in passing this judgment on him—in terms reminiscent of Marlow's judgment of Jim: "He cannot act or exist without idealizing every simple feeling, desire, or achievement. He could not believe his own motives if he did not make them first a part of some fairy-tale. The earth is not quite good enough for him, I fear" (237–38). It is not Martin Decoud's "egotism" that speaks, but his instinctive intelligence.

In Norman Sherry's exhaustive *Conrad's Western World* —a masterpiece of that kind of scholarship which lists facts, "sources," and endless detailed references to the author's personal experiences—one learns that "Martin Decoud" had a historical origin; or, at any rate, that Conrad had access to historical works that mentioned a "Decoud" (along, of course, with other characters in *Nostromo*). In Sherry, and in other biographical works, one learns as well the interesting fact that Conrad had evidently attempted suicide himself, as a young man; like Martin Decoud, he shot himself in the chest.[9] There are a number of characters in Conrad who commit suicide or who arrange to be killed, in suicidal gestures elevated to the level of heroism, like Jim's, and in *Lord Jim* there is the enigmatic Captain Brierly, a king among sea-captains, with a perfect record and an enviable reputation among other, lesser men. The strain to be perfect, to continue to be perfect, is deathly, for "perfection" of the kind Brierly had achieved leaves one no human possibilities, nowhere to go except down; because Brierly is intelligent enough to realize that Jim's act of assumed cowardice might someday be his own, he decides to kill himself at once, before his record is blemished. The more one allows one's ego to be inflated, by the opinions of others, the more attractive does dissolution become, for each extreme hungers for its antithesis, and the rigidity of perfection inspires an irresistible fascination for utter collapse, chaos, dissolution. Even more, it is the non-human or anti-human "indifference of *things*" (as opposed to the claustrophobic and confining mirror-house of the interest of human beings) that appeals. It is likely that Conrad identified strongly with those of his heroes who did commit suicide. In most cases the "suicide" of a fictional character is probably a cleansing, therapeutic event in the psychic life of the creator, for the symbolic exorcism of deathly or ruinous character traits allows the creator not only the energetic pleasure of destroying something that needs to be destroyed, but the perspective, the detachment, that follows this destruction. That is,

Marlow and Captain Jones can discuss the formidable Brierly's suicide and, between the two of them, hit upon a conclusion of "amazing profundity"—that Brierly was an ego-maniac, and that they are not. Marlow can speak of Jim and of Kurtz from the advantageous position of simply having outlived them, and though he is respectfully ambivalent in his judgment on them, one is left with the feeling that they represented diseased or exaggerated or unworkable values: it is not "good" that these human beings died, but "good" that the values they represent are seen as deathly. When author-subject identification is almost total, and the death-wishes of the protagonist are taken with utmost seriousness, of course there is no exorcism—there is no depersonalization of the self into its various components, and consequently no detached position from which the melodrama of psychological components can be witnessed. In such cases the death-wishes probably remain in the writer and will simply attack him from another, newer angle, until the point at which—quite literally—either the intellectual principle or the emotional principle triumphs. One aspires to order and detachment, the other to a surrender to feeling—and "feeling," in the neurotic, cannot be trusted as being in the best interests of the personality. Art parallels the symbolic cleansing actions of dreams, rather than the literal actions of "reality," for it deals with psychic contents imagined as actual, existing characters or events, deals with them both in play and in utter seriousness, and allows for that detachment from the hot, emotional, short-sighted concerns of the immediate personality. Thus it is highly therapeutic precisely when it is highly artistic: genuine tragedy works to shake and frighten and exalt, forced tragedy works to undermine the very structure of society.

The "tragedy" of *Nostromo* reduces human beings to mere units in the *farce macabre* of history, and one would certainly not question this vision—it is as convincing as its antithesis, the religious view which holds that social and political events matter very little in the individual's life; but the novel, in dealing with so broad and dehumanizing a theme, seeks nevertheless to place the burden of responsibility and guilt *inside* the individual. Martin Decoud goes to his death because he has had the bad luck to be involved in political activity, not because there is anything deluded or wicked in him. It is not a moral or aesthetic necessity that he die, as it is that Kurtz be unmasked and destroyed, or the various luckless creatures of *The Secret Agent* and *Under Western Eyes* be effectively defused—from the point of view of the conservative Conrad. Killing Decoud is simply killing Decoud—killing the values and insights and capacity for adoration of the female that Decoud represents.[10] It is not an exorcism, merely an erasure, which leaves *Nostromo* rudderless, diffuse, packed with brute events, a serio-comic extravaganza which could conceivably go on forever—without a center of consciousness that defines it, that we can accept without irony. The demystified Nostromo gains in insight as the novel concludes, but is not sufficiently complex to function in the way that Decoud does; he is not tragic, merely unfortunate. And Dr. Monygham, with his almost complete negation of human values, is far too cynical, and too broken, to form a center of consciousness by which the novel's final ironies are interpreted. Surprisingly, it is Mrs. Gould who achieves a considerable insight—in seeing the position she has had to play in regard to her husband, whom she has nevertheless lost, emotionally; but her sudden cynicism, her almost sadistic "honesty" in telling Nostromo's beloved, Giselle, that he would have "forgotten you for this treasure" (626) suggests that she too has been corrupted by the silver of the mine—or by her own unreal role in Sulaco. Her fairy-like *persona* crumples, her exaggerated ideals are rejected, and she becomes a fit mate, spiritually, for the crippled Monygham.

Decoud dies for love, however misguided it might strike us; he dies for patriotic reasons, however anxiously he chooses to reject the very conception of patriotism. The tragic possibilities of his position are not adequately developed, and *Nostromo,* though not a tragic work in any conventional sense of the term, might be viewed as an attempt to express in novel form the "poetic art" of tragedy Schopenhauer valued so highly because it dramatized symbolically the conflict of will with itself—in which the individual counts as very little, and his suffering is no more than an expression of the general want and drudgery of existence. In such a universe, action necessarily leads to destruction; but the refusal to act leads to destruction as well. Can meaningful tragedy be written out of a nihilistic vision? Conrad makes the attempt in deadly earnest in *Victory,* and it is questionable that he succeeds, for a post-Victorian age. Disguised as "sacrifice," "love," and "ideals," disguised, even, as theatrical "evil," the basic elements of Conrad's universe simply reassemble themselves for the routine, hopeless struggle, in which all must lose. In "The Platonic Idea as the Object of Art," in *The World as Will and Idea,* Schopenhauer praises tragedy because of its power in expressing a significant "hint" of the real nature of things. His vision is essentially that of Conrad, in *Nostromo* more than any other work. The heroic individual is literally drowned in the flood of existence:

> It is one and the same will that lives and appears in them all, but whose phenomena fight against each other and destroy one another. In one individual it appears powerfully, in another more weakly; in one more subject to reason . . . in another less so, till at last, in some single case, this knowledge, purified and heightened by suffering itself, reaches the point at which the phenomenon, the veil of Maya, no longer deceives it. It sees through the form of the phenomenon, the *principium individuationis*. . . .

THE SILVER OF THE MINE: "MATERIAL INTERESTS" AS SPIRITUAL DESIRE

Conrad effectively uses civil war in Costaguana to dramatize various kinds of division: that between the hapless, victimized poor and those who exploit them, that between the Costaguaneros and the Europeans and Americans who are eager to "develop" the country, that between the intellectual and the rest of mankind, that between male and female, female and male, one's image and one's identity, one's fragmented historical personality and one's integrated ahistorical self.

Like the ivory of *Heart of Darkness*, the silver of the San Tomé mine casts a kind of enchantment over the land. Men are bewitched by it, or seduced into desperate action on its behalf; like the accursed gringos, drawn to treasure out onto the peninsula of Azuera and thereby trapped under the "fatal spell of their success," they lose their souls—are paradoxically rich and yet starving. In one sense *Nostromo* is a painfully prophetic work, a kind of parable of our times. It looks ahead to the mid-twentieth century, to a future in which "material interests" have dominated world politics, and the justice of which Charles Gould so complacently speaks is nowhere to be found. Representatives of "advanced" nations are eager to finance revolution in Costaguana, as they are always eager to finance revolutions in certain countries; imperialism is given the code-name of a "future." As an Englishman assures Mrs. Gould, "You shall have more steamers, a railway, a telegraph-cable—a future in the great world which is worth infinitely more than any amount of ecclesiastical past" (39). Only through the exploitation of the San Tomé mine can this future be achieved, and only through civil war can the mine be secured for the interests of foreign developers. If the silver of the mine is the novel's true subject, in a symbolic sense, the motif of civil war is the novel's dominating theme. On many levels mankind is divided against itself, and what phenomenon inspires it to this self-destructive division, except the quest for material treasure? (Conrad is careful to record the fact that prerevolutionary Costaguana was a corrupt, insanely brutal state; nevertheless, he states clearly his belief that the new, "enlightened" Occidental Republic is no improvement, and more political turbulence is promised at the novel's conclusion.) Dr. Monygham makes the connection between the relatively innocent, idealistic opening of the mine, years before, and the continued unrest; and makes this ominous prediction:

> There is no peace and rest in the development of material interests. They have their law and their justice. But it is founded on expediency, and is inhuman; it is without rectitude, without the continuity and the force that can be found only in moral principle. . . . The time approaches when all that the Gould Concession stands for shall weigh as heavily upon the people as the barbarism, cruelty, and misrule of a few years back. [571]

Significantly, hearing this, Mrs. Gould cries out "as if hurt in the most sensitive place of her soul." Prophetic also, rather amazingly accurate, is Conrad's ironic view of American imperialism, always disguised as something idealistic or even "religious." The millionaire Holroyd, nicely named, and with an ancestry that calls to mind the Melting Pot of democracy—German, Scotch, French, English, Danish—tells Mrs. Gould that the mistakes of old Europe are not likely to be the mistakes of America. This newer, more robust, and more cunning land, which synthesizes the Puritan temperament and the insatiable imagination of the conqueror, knows how to deal with the isolated, impoverished Costaguana. Holroyd says modestly:

> We in this country know just about enough to keep indoors when it rains. We can sit and watch. Of course, some day we shall step in. We are bound to. But there's no hurry. Time itself has got to wait upon the greatest country in the whole of God's universe. We shall be giving the word for everything—industry, trade, law, journalism, art, politics, and religion, from Cape Horn clear over to Smith's Sound, and beyond, too, if anything worth taking hold of turns up at the North Pole. And then we shall have the leisure to take in hand the outlying islands and continents of the earth. We shall run the world's business whether the world likes it or not. The world can't help it —and neither can we, I guess. [85]

This collective destiny necessarily ignores or corrupts or destroys the individual, who cannot resist participating in the delusions of this culture. So Nostromo, who has allowed the crowd to define his worth, and who has been in the employ of those Europeans who know best how to value him, suffers a kind of spiritual death when his image of himself is snatched from him; the world of extraverted values, of material, measurable interests, has corrupted him. As he is presented as a "Man of the People," Conrad suggests that the "People" too are subject to this dementia, the more so as their ecclesiastical tradition is lost. The material interests mean to stabilize the government of Costaguana and bring about a better justice have changed everyone, but along the lines of the material rather than the spiritual; so the workers of the San Tomé mine, the earliest laborers in Gould's hire, were actually willing to defend the mine and to march upon the captured town (thereby saving it for the foreign investors), but at the present time they assuredly would not; they would never risk their lives for the Gould Concession. The people have changed, have been changed irreparably. And as the Occidental Republic is one of the "treasure houses" of the world, it is possible to see the transformation of the world, writ small, in the allegorical drama of *Nos-*

tromo. At first victimized, in their innocence, the people then acquire some of the traits of their exploiters. They must learn to be self-serving, or they will not survive. Even so, the survival of Nostromo is predicated upon his spiritual destruction; his actual death is an anti-climax.

Valuable as *Nostromo* is in prophesying the inevitable effects of material interests and their exploitation, in ostensibly backward states, it is even more rewarding if considered as an allegorical work, one which is concerned with dramatizing the plight of the soul; like *Heart of Darkness*, which it resembles closely in its poetic approximation of inner states of mind to exterior circumstances, it centers its narrative on a single image (in this case the silver of the mine) and a single repetitive action (the "civil war" on various levels).

The gringos of Conrad's *exemplum* are legendary and mythical, but their spiritual damnation is shared by the novel's presumably more human individuals. Conrad presents his theme in imagistic terms, then continues with the dense, clotted exposition; he touches upon the mine, the silver, the treasure, but is careful, always, to continue with the narrative, the entangled, exasperating history, in which causal relationships seem at times defeated, and individual struggle exposed as futile. The novel is symbolic, and then aggressively realistic; it smothers us with detail, and then draws back to make certain psychological analogies between the silver and the interior delusions that guide most of the novel's characters. Naturalistic in the fundamental method by which its narrative is unfolded, it is simultaneously—or intermittently—symbolist in intention. Now one vision dominates; now another. The awkward ending, which strains to effect a closure unrelated to the preceding narrative—the histrionic, faithful Linda in the lighthouse, crying of her love for Nostromo out into the darkness—is puzzling on a naturalistic level, and does not work very smoothly as art, but reveals its meaning if explored as a poetic sequence.

Though the fallen hero, Nostromo, is lost to her, the eternally vigilant, devoted female, Linda, presides over the darkness—shines the light of the Great Isabel's lighthouse over the lost treasure of the mine. Male and female cannot be united, but there is a powerful, tragic grandeur—in Conrad's imagination, at least—in the fact that the treasure, invisible and secret, will be kept by the woman, intact and incorruptible at last. All action is harmful, as we have been warned; but there is the possibility of transcendence by the rejection of action, after the tragic death of the hero, by the surviving beloved. *She*—whether Linda, or Antonia, or the transparently allegorical figure of Kurtz's Intended—will be incorruptible because fleshless, pure spirit, the defiant custodian of an outworn but heroic truth.

In myths, legends, and fairy tales, "treasure" is always symbolic; it speaks of inner visions, of the hidden, and takes its convenient images from the exterior world. A psychoanalytic reading of *Nostromo* would surely not overlook the sexual implications of the silver of the mine, the treasure of the womb, the commodity over which Costaguana's men are willing to die, as rivals for its riches; Charles Gould, the King of Sulaco himself, is called "the poor boy" by his own wife—and Conrad speaks of his subtle "infidelity," his eventual shifting of emotional interest from his wife to the mine, that more primeval center for his consciousness. Freudian theory would have it that most, if not all, of civilization results from the thwarting of aggressive and incestuous wishes, and the necessary sublimation of infantile urges into "adult" projects. Gould's "adulthood" allows him to be both infantile and kingly; he is even, by deliberately disobeying his dead father's wishes, a small god, defying fate. In such terms, Martin Decoud, making off with the silver, must be punished because he has dared—however indirectly—to commit an incestuous feat, along with his brother-rival, Nostromo. But a larger, more spacious psychological theory would move away from the limitations of the biological, homeostatic norm, to suggest that the dominant image of the silver is suggestive of spiritual, rather than physical, unions. So the innumerable Biblical treasures, the sudden "radiances" of mystical literature, and the actual precious metals or gems—like the Pearl—are attempts to express, in poetic terms, ineffable truths about man's relationship with the deeper aspects of his own nature. It is significant that in the West, where spirit and flesh have customarily been separated, images of divine experience are usually inanimate; in the East, in such religions as Hinduism, where spirit and flesh are imagined as one, the symbol of divine experience can be an image so aggressively physical as that of the erotically joined male and female, which symbolizes the union of "masculine" and "feminine" in the personality, and the "bliss" of the awakened, integrated soul, itself asexual, neuter, transcendent. In the West, religious imagery is often interpreted as disguised sexual imagery; in the East, sexual imagery is meant to be interpreted in terms of religious experience. Psychoanalytic interpretation is often helpful, when art is approached as a neurotic achievement, or as an achievement closely bound up with the "neurosis" of our culture, but its limitations are obvious.

Nostromo is one of those odd, captivating works in which, possibly because of the author's inability or refusal to make conscious everything that is really involved, satisfactory endings to various problematic human relationships cannot be achieved. If there is a tragedy here, it is the terrible, helpless diffusion of energies—the fear of union—the almost superstitious awe with which the sexes view one another, since they have exaggerated, unconsciously, the "powers" inherent in the other. Woman, for Conrad, is always a mystery. As Marlow says, women are altogether different from men; they live in a world of their own illusions, and should not be enlightened, for civilization itself depends upon their blindness and their continued acting-out of the traditional role of the "female." If they are literally fallen women, like Lena of *Victory,* they can redeem themselves—and their lovers—by acts of personal sacrifice, in which love itself is extinguished; the supreme gesture being a kind of suicidal plunge, after which the lover, like Heyst, can commit actual suicide. In this way the sexes are united—and there is no danger of the species being reproduced. Jim's "Jewel," whom he does not so much betray as forget, in his eagerness to die a hero, is ironically named; it is not she Jim will really mate with, but death. Her "value" is only rhetorical, and her fundamental distrust of the male of the species seems to anticipate, if not unconsciously provoke, Jim's betrayal. In *Nostromo,* the silver of the mine is nearly always peripheral to some masculine ideal, like Gould's dream of justice, or Nostromo's assuaging of his damaged ego, or Decoud's love for Antonia; but it symbolizes not the possibility of these ideals, but their utter defeat. There cannot be union, there cannot be a synthesis of the ideal and the real, the eternal and the temporal, the ahistorical and the historical. The "silver" is an obsession because the true nature of man's struggle has not been made conscious; lacking an awareness of the need for spiritual unity in his soul, man naturally projects value outward, into the extraverted, material world, which he cannot hope to conquer. Like Sotillo, he goes mad seeking riches in a literal sense. "The Kingdom of God is within" is a symbolic expression of the psychological possibility of treasure within, value independent of history, and of others' opinions, within the sphere of the individual himself. The existence of any "God" is not an issue; even metaphysicians project their own fancies into systems; but the experience of the integration of the split, warring parts of the psyche, which in mystical literature is often described as the "union" of ego and soul, or "masculine" and "feminine" inclinations, is usually the goal toward which literary heroes or heroines strive, however secular or even banal (as in Jane Austen's novels, where the legal "wedding" is the central event of one's life) their individual goals might seem.

Most of Conrad's characters are paired off, like dancing partners if not loving partners, but something falls between them—the shadow of the ivory, or the silver; unity is desired, at least philosophically, but the serio-comic episodes of life defeat it. But is union desired?—in its literal, human aspects? The female is always there, as an ideal presence, fairy-like, virginal, "godly" only in the minor and debased sense in which the Virgin Mary is "godly"; but she presides from afar or from a height, conferring her blessing upon masculine activity. Typically, Mrs. Gould watches the silver escort pass through Sulaco on its way to the harbor, in one of Conrad's innumerable "visually" poetic scenes:

> In the whole sunlit range of empty balconies along the street only one white figure would be visible high above the clear pavement . . . a mass of heavy fair hair twisted up . . . and a lot of lace about the neck of her muslin wrapper. With a smile to her husband's quick, upward glance, she would watch the whole thing stream past below her feet with an orderly uproar. . . . [127]

Though he is not involved in the silver of the mine in any literal sense, she is spiritually necessary; she lends her support to this enterprising hero, and thereby confirms the split, the compulsive overestimation of the material world, which will account for the tragedy of the future. Conrad's work is generally depressing because, apart from scattered and unconvincing statements of optimism—of the need for human solidarity—there is very little awareness, on the part of the author, of the genuine need for *literal* unities (which might bring with them, as they do in visionary writers like Lawrence and Whitman, the possibility of spiritual unities). Lena and Heyst are not literally united, as lovers or even as friends; they are only rhetorically united, by Conrad's will. No rush of sympathy or warmth can be allowed between Mrs. Gould and Dr. Monygham, for their emotional mating cannot be given the slightest physical cast; Emilia Gould is an "ideal" wife, in the Victorian mold, which is to say a stunted and sexless human being. Decoud's adventure is doomed, for he has the audacity to admit that it is passion that motivates him, not patriotism; something must divide him from the fulfillment of his desire, his union with Antonia. Nostromo clings to Giselle as to life itself—which indeed she is, in a sense—but the silver will come between them, Nostromo must die, in a drama of phenomena in which mere impulses combat one another, the disguised Will men take as World.[11]

Conrad's characters are trapped in contradictory and mutually exclusive imperatives. They must exist in time and yet they must be incorruptible; they must affirm human community, and yet they must be careful of any real intimacies, and direct human bonds. The splitting of the ideal and the real, like the splitting of the "spiritual" and the "flesh," has been one of the curses of Western consciousness, and Conrad is one of the West's supreme imaginative artists, at least for our modern era, which has denied supernatural extremes of Diety and Devil, but has replaced them with intermediary, relativistic temptations. "Duty" as a working substitute for "Deity" is a standard Victorian arrangement, and Conrad is very Victorian in his deepest soul. Consequently, any force that questions "Duty"—like the disrupting forces of emotional or spiritual unrest—must be denied, seen as mutinous, dangerous. Conrad felt no "exaltation" upon the completion of *Nostromo,* perhaps because he sensed how the novel works to destroy, rather than to affirm, any human values whatsoever. Not even "evil"—or the pleasure of motiveless, destructive evil—is recognized as a valid human experience, as it is in Dostoyevsky. "One must work to some end," Charles Gould says vaguely, and it seems not to matter that the end is a trivial, material one: this desultoriness in the face of quite real, literal events assures tragic consequences, if not tragic grandeur.

Because the "treasure" of the soul is bound up with the realization of the male's need for unity with the female, and the female is distrusted, or acceptable only in a merely spiritual sense, man is trapped in time; he cannot transcend history. To awaken feminine contents in his own psyche would be a softening of his masculine "Duty," which cannot be risked—better death, better dissolution, than such a surrender to non-heroic ideals. What is humankind's most instinctive gesture toward union but a lamentable swallowing of that "barbed hook," which brings innumerable doomed generations out of the Void?

Notes

1. By following Stein's famous imperative—"in the destructive element immerse"—Conrad's characters invariably destroy themselves; and if nothing redeems mankind's sordid activities except the "idea," how can one continue to participate in these activities when the "idea" is exposed as "illusion"? What Paul Kirschner calls a "limitation in Conrad's psychology" (". . . his refusal to grant validity or grandeur to dreams not associated with personal spirit. . . . That a man might be sincerely and legitimately moved by a dream of social justice or reform did not seem to occur to Conrad," in *The Psychologist as Artist*, Edinburgh, 1968, p. 286) is not inevitably a limitation, but the expression of a violently pessimistic and nihilistic nature.

2. Of course Conrad is Victorian—in part; and he speaks for certain Victorian ideals. But since he is savagely critical, elsewhere, of Victorian ethics, it can be assumed that he is speaking for Joseph Conrad himself in such lines as these:

> A woman with a masculine mind is not a being of superior efficiency; she is simply a phenomenon of imperfect differentiation —interestingly barren and without importance. [Mrs. Gould's] intelligence being feminine led her to achieve the conquest of Sulaco. . . . She could converse charmingly, but was not talkative. The wisdom of the heart having no concern with the erection or demolition of theories . . . has no random words at its commands. . . . A woman's true tenderness, like the true virility of man, is expressed in action of a conquering kind. [73–74, *Nostromo*]

Femininity, then, like masculinity, is valued for its ability to conquer; woman too is a warrior, and indeed has no value otherwise.

3. "Mr. Jones," the gentleman, like "Gentleman Brown" of *Lord Jim,* is death-in-life, a walking cadaver. Conrad describes him as "sunken-eyed," "hollow-voiced," a being with a "used-up, weary, depraved distinction." He is a "spectre" and his voice is dead, as if coming from the "bottom of a well." At the same time, rather like Faulkner's Popeye, in *Sanctuary* (a work that is remarkably "Conradian"), he is the pivot of the plot, the fantastic creature who sets in motion the ostensibly "normal" characters—and brings about their defeats. "Mr. Jones' " protestations, his hatred of all women, make him, curiously, one of the very few characters in all of Conrad who seems believable as a sexual human being. What in the others is mere rhetoric (Heyst's love for Lena, for instance) is, at least, passion of a kind in "Mr. Jones." And he has a sense of humor–another villainous trait.

4. In ironic contrast to the more public Conrad, eager to assert a faith in the "simple verities," is the Conrad revealed in a letter to R. B. Cunninghame Graham:

> The attitude of cold unconcern is the only reasonable one. Of course, reason is hateful—but why? Because it demonstrates (to those who have courage) that we, living, are out of life—utterly out of it. The mysteries of a universe made of drops of fire and clods of mud do not concern us in the least. The fate of humanity condemned ultimately to perish from cold is not worth troubling about. . . . [Quoted in Robert Penn Warren's Introduction to the Modern Library edition of *Nostromo,* p. xx.]

Such passages suggest that Conrad, the novelist, is really more sympathetic with his skeptical characters, like Heyst's father, than he is with his ostensible heroes.

5. It is not at all clear why Conrad retained the title *Nostromo*, which he seems to have known was misleading; perversely, he insists that the "silver of the mine" is the real subject:

> . . . I will take the liberty to point out that *Nostromo* [sic] has never been intended for the hero of the Tale of the Seaboard. Silver is the pivot of the moral and material events, affecting the lives of everybody in the tale. That this was my deliberate purpose there can be no doubt. I struck the first note of intention in the unusual form which I gave to the title of the First Part, by calling it "The Silver of the Mine," and by telling of the enchanted treasure . . . which, strictly speaking, has nothing to do with the rest of the novel. The world "silver" occurs almost at the very beginning of the story proper, and I took care to introduce it in the very last paragraph, which would perhaps have been better without the phrase which contains the key word. [from a letter to Ernst Bendz, quoted in Frederick R. Karl, *A Reader's Guide to Joseph Conrad* (New York, 1965), pp. 155–56.]

6. See Harry Marten, "Conrad's Skeptic Reconsidered: A Study of Martin Decoud," in *Nineteenth Century Fiction*, Vol. 27. All Decoud's actions, all his roles—as skeptic, lover, part-patriot—can be disregarded because the man is nothing more than a supreme egotist who wants only to feel important.

7. In *Joseph Conrad* (London, 1968), p. 131.

8. The conservative temperament always praises "strong" and "mute" workers; it singles out for displeasure such rebels as James Wait and Donkin, who bring to the sleepwalking world of such tiny societies as the *Narcissus* a raucous, questioning attitude that will, if not smothered, upset the status quo. Conrad's difficult position is that, as the imaginative artist, he knows how James Wait and Donkin—as well as Kurtz, Mr. Jones, Gentleman Brown, and even such colorful minor "villains" as Jim's Captain on the *Patna*, and the various Monteros and Sotillos of *Nostromo*—are the very breath of life itself, the channels by which motion, change, *drama* are introduced into the world of the narrative. Tragedy allows such people onstage, allows them to destroy the presumably "heroic"—and then punishes them, and their values, in order to keep chaos outside the world and aesthetic boundaries of the work of art. Tragedy is basically a conservative art-form. But the writer of tragedy can be neither conservative nor rebellious; he must have sympathy with every point of view engaged in the

drama, or his work will fail. When Conrad is polemical he is often, curiously enough, praising and inflating values not his own; as in this passage from *The Nigger of the "Narcissus,"* where the old reliable seaman, Singleton, is apotheosized:

> His generation lived inarticulate and indispensable, without knowing the sweetness of affections or the refuge of a home—and died free from the dark menace of a narrow grave. They were the everlasting children of the mysterious sea. Their successors are the grown-up children of a discontented earth. They are less naughty, but less innocent; less profane, but perhaps also less believing; and if they have learned how to speak they have also learned how to whine. But the others were strong and mute; they were effaced, bowed and enduring, like stone caryatides that hold up in the night the lighted halls of a resplendent and glorious edifice. [21, Vintage edition]

9. Conrad's uncle wrote of the suicide attempt, which was evidently a result of debt:

> . . . One fine evening [Conrad] invites his friend the creditor to tea, and before his arrival attempts to take his life with a revolver. . . . The bullet goes *durch und durch* near his heart without damaging any vital organ. . . .

(Quoted in various books, among them Norman Sherry's *Conrad's Western World* (London and New York, 1971), p. 168.)

10. It could be argued, however, that it is Decoud's adoration of Antonia, his eagerness to distort his beliefs to fit hers, that assures his death. Only so long as he is a "patriot" does Antonia love him; so he becomes a hack journalist, a propagandist, he compromises his integrity, betrays his own nature. In such an interpretation, Antonia is the beautiful "Other" (to use Conrad's expression in his preface to *Nostromo*) who lures men to their destruction. But Conrad seems not to have had such a deep, passionate emotional attraction for or repulsion against women. Decoud's schoolboy enthusiasm regarding his love for Antonia may well be Conrad's own, since his "first love," the young Polish patriot, did indeed belong to Conrad's schooldays. Here is Decoud speaking quite boyishly:

> I am not deceiving myself about my motives. She (i.e., Antonia) won't leave Sulaco for my sake, therefore Sulaco must leave the rest of the republic to its fate. Nothing could be clearer than that. . . . I cannot part with Antonia, therefore

> the one and indivisible republic of Costaguana must be made to part with its western province. [238]

11. In such drama, masculine elevation of the "idea" is always threatened by feminine temptation; the female of the species bears the burden of being closer to "Nature," perhaps even allied with the undifferentiated process of time. The typical man is, then, in a double-bind situation. He requires the female for the completion of his masculinity, for the fulfillment of his life; but if he surrenders to her, he is surrendering to a presumably brute and ignoble aspect of his own nature. Woman's role, in Schopenhauer's imagination, is to do nothing more and nothing less than to continue to "seduce" men, in order to perpetuate the race. The heroic is constantly defeated by the emotional. Conrad's views are never so clearly stated as Schopenhauer's, but they are remarkably similar. See Schopenhauer generally, and in such passages as this one from "The Metaphysics of the Love of the Sexes," in *The World as Will and Idea:*

> Through [a surrender to erotic love] man shows that the species lies closer to him than the individual. . . . Why does the lover hang with complete abandonment on the eyes of his chosen one, and is ready to make every sacrifice for her? Because it is his immortal part that longs after her; while it is only his mortal part that desires everything else. . . . In the midst of the tumult we see the glances of two lovers meet longingly; yet why so secretly, fearfully, and stealthily? Because these lovers are the traitors who seek to perpetuate the whole want and drudgery [of human life], which would otherwise speedily reach an end.

"In the Fifth Act": The Art of the English and Scottish Traditional Ballads

Mony a one for him makes mane,. .
But nane sall ken where he is gane;
O'er his white banes, when they are bare,
The wind sall blaw for evermair.
"The Twa Corbies"

We are each haunted by different things, and we do not choose the presences, angelic or demonic, that haunt us: rather, we know ourselves chosen. There are a number of ballads, and an even greater number of isolated lines from ballads, that have imprinted themselves irrevocably on my imagination. Why is it that the four concluding lines of "The Twa Corbies" should mean so much to me? And the last three stanzas, dirge-like, unspeakably beautiful, of "Sir Patrick Spens"? If I am alone and I become aware of myself humming or singing under my breath, why is it likely to be "Barbara Allen"? Questions bombard: Are there others in whom these ballads float upward into consciousness at the oddest, least plausible times? What does it mean to say that one is "haunted" by anything, let alone literary works? There are many of us, I suspect, in whom random lines from Shakespeare have sunk as deeply, and have taken on as powerful and as intimate an authority, as any experience from what we are accustomed to call "real" life.

Ballads tell stories, and stories, after all, exert an irresistible appeal upon us all: perhaps this accounts for the mystery. The rhythms of narrative, the very phenomenon of narrative itself—these strike me as fascinating; surely they correspond to some deep rhythms of our own, unconscious, even physical, in origin; and a highly self-conscious Modernist art that rejects "plot" is in danger of rejecting the very origins of art's impulse. But the ballads strike deeper; it is not for their stories alone that they compel us and have the power to frighten us; there is very little story, in fact, in "The Twa Corbies." Thomas Mann in *Dr. Faustus,* in presenting the psychological picture of Adrian Leverkühn's hometown, Kaisersaschern, speaks of "the folk" with both disdain and awe, and a kind of frightened reverence as well—"this old, folkish layer that survives in us all"—the stamp of "old-world, underground neurosis" that, to a friend of the Enlightenment (for so Mann saw himself, and his reader) has always something anachronistic and alarming about it. The ballads possess this flavor, this tone, this primitive authority: I see them as expressions of the human imagination on a very nearly unconscious level, the kind of art, "primitive" or otherwise, that deals with the magical, the supernatural, the impossible—in short, the fantasies of the Unconscious—as if they were quite natural; as if, in fact, they belonged to the day-side and not the night-side of our existence.

The ballads are child-like in their simplicity, their intensity, their use of sing-song repetitions, and their presentation of events without apparent symbolic or spiritual meaning. This is how it happened, these are the steps by which it happened—that's all. Nothing is left over. Nothing transcends narrative. Of course they are among our oldest forms of literature and their origins cannot be traced. They are almost without exception "tragic" in the most common sense of that word—that is, not trivial, not "comic"—and even when they are clearly meant to constitute a recording of historical fact, that history is usually tragic. An oral art—the ballads *are* art—a communal art, they were shaped by innumerable artists, and there is a sense in which they belong to the species, to us all, in a way that our most highly refined and individualized art-works—Joyce's *Ulysses* and *Finnegans Wake,* for example—cannot.

When Thomas Gray said of the ballads that they commonly begin "in the fifth act" he was suggesting that their emotional appeal is a result of their accumulative narrative structure; they have the form, though Gray of course does not say this, of the modern short story which begins near its end, its chronological end, and often doubles back on itself in order to bring the reader into the emotional nexus of the story. Most of the ballads are simple stories of action. Hearing them sung—"Johnnie O'Breadisley," for example—one is carried along quite willingly; what happens *is* all that matters; moral commentary seems superfluous. Then there is "Fair Margaret and Sweet William" and its ubiquitous variants (including "Barbara Allen"), in which one lover dies for love, and is soon joined by the other: and above their graves grow a brier and a rose which join, as they must, in "a true lover's knot." Here, commentary is implicit; something is being said about the nature of love, or the nature of love as we might wish it to be. And there is "Clerk Colvill" or "Lady Alice" and its variants (including "George Collins"), in which a young man, evidently in excellent health, is told by a "fair maid" who belongs to an order of reality different from his own that his life will soon end: he embraces the young woman nevertheless, and kisses her, and goes home to his father's house to die. (And in most versions his human sweetheart dies shortly afterward.) The ballad is mesmerizing, the quick jerky incidents as compelling as incidents in a dream, offered without exposition; one knows the story did not happen in this way, could not have happened in this way—and yet, emotionally, psychologically, perhaps it did. Commentary of some kind, even critical analysis, is inviting; one wants to leap centuries ahead to Keats's "La Belle Dame sans Merci": a highly conscious and perhaps too fastidiously groomed variant of the ballad which, compared with the ballad, is not inevitably superior. For instance, Keats:

I met a lady in the meads,
Full beautiful—a faery's child,
Her hair was long, her foot was light,
And her eyes were wild.

I made a garland for her head,
And bracelets too, and fragrant zone;
She look'd at me as she did love,
And made sweet moan.

And the variant "George Collins" (Child 85):[1]

George Collins walked out one May morning,
When May was all in bloom.
There he espied a pretty fair maid,
A-washing her marble stone.

She whooped and she hollered, she highered her voice,
She held up her lilywhite hand.
"Come hither to me," George Collins," she said,
"For your life shan't last you long."

(One is pleased to note that though the water faery is burdened with the cliché of a "lilywhite hand" she is nonetheless robust enough to "whoop" and "holler" and "higher" her voice—the last, a rather wonderful word we seem to have lost.) If the faery is denied her status in the real world, as a psychological interpretation insists, it still isn't clear what she represents. Is she, in Jungian terms, a false Anima—that is, a repressed and hence destructive side of the young man himself, his "female" self which he has failed to allow into consciousness?—for the psyche insists upon freedom, the liberation of instincts, the balance between "male" and "female" in us all. Yet a close reading of the ballad suggests that the faery simply tells the mortal man that he will die soon; a kiss from her does not *cause* his death. So the psychological explanation, though it is certainly appealing, really does not work here—as it works in Keats, where the narrator, the pale knight-at-arms, is lulled to sleep by the faery and dreams of "pale kings" and princes and warriors who, like the knight himself, are in the thrall of "La Belle Dame sans Merci." Surely it is a commonplace in clinical psychiatry—that a man is thwarted in his masculinity because of an adolescent or even infantile fixation upon the "female" which is a part of his own soul, but denied; and hence murderously powerful. But the original ballad cannot be so analyzed. There is something in it that remains inexplicable, haunting. Perhaps, one halfway thinks, the faery was *real?*

In another nearly ubiquitous ballad, "The Prickly Bush" (Child 95) or "The Maid Freed from the Gallows," a young woman is to be hanged for an undisclosed crime, but she will be freed, evidently, if someone brings the hangman enough gold. (The story situation, with its melodrama, its suspense, has made it tremendously attractive—it is found in almost every European country as well as in the English-speaking nations, and has even been transposed into a children's game; ballad scholars say that it is the best known of the traditional ballads among American and West Indian blacks—though the protagonist is usually not a young woman but a young man.) The version recorded in Child begins bluntly, boldly, without explanation:

> "O hangman, hold your hand," she cried,
> "O hold it for a while,

For I can see my own dear father,
Coming over the yonder stile."

"O father, have you brought me gold,
Or will you set me free,
Or be you come to see me hung
All on this high gallows tree?"

But the father replies, unpredictably, astonishingly:

"No, I have not brought thee gold,
And I'll not set thee free,
But I am come to see thee hung,
All on this high gallows tree."

Then the refrain, the chorus, which is repeated after father, mother, and brother deny her:

"O the prickly bush, the prickly bush,
It pricked my heart full sore,
And if I ever get out of the prickly bush,
I'll never get in no more."

At last, the young woman's lover appears: and he has brought gold to free her. The ballad ends with a rueful: "But now I'm out of this prickly bush,/And I'll never get in no more." As a drama "The Prickly Bush" is unsurpassable, and though one false hope follows another the listener is always attentive; even when one knows the ballad quite well, he is *always* caught up in the story. But the "prickly bush" itself—what does it mean? Difficulty with the law? Difficulty in love? (The "prickly bush" is a frequent symbol of disastrous love in British folk-songs; a young woman, betrayed by her lover, may go into the forest to bear her illegitimate child alone, in great agony, pressing up against a thorn bush—presumably to help her endure the greater pains of childbirth.) The narrative completes itself with the freeing of the young woman by her lover, but it does not seem to us, with our modern sensibilities, *complete;* too many questions go unanswered. Is the crime for which a young woman might have been hanged and because of which her family might have publicly rejected her most plausibly the crime of infanticide? In that case one would want to know why her lover abandoned her previously; but of course the ballad offers no explanations, not even any suggestive clues.

The anonymous creators of the ballads, it is extremely interesting to note, are not burdened with gender like the rest of us: male and female blend, become one, so that the typical "voice" of the ballad is asexual as well as omniscient. If, in reading Shakespeare or Joyce or Lawrence, we marvel at the ability of a man to take on the consciousness of a woman, at least in speech, in words, we are always aware that the author is a man, though certainly a man of "genius." The ballads, however, have the advantage of being folk-art, and therefore not the work of individuals. Sung by both men and women, they acquire the smoothness, and also the hardness, of a much-polished stone; though their observations of the grotesque inequities between the lot of men and the lot of women—not only in the dirge-like songs of young women who murder their illegitimate children, but in such hearty, good-natured songs as "The Trooper and the Maid" (Child 299) and its innumerable variants—carry with them, typically, no outrage, not even the mildest condemnation, the very presentation of certain events, and the frequency with which those events are recorded, constitute a judgment. Never more than implicit, however: the ballad-singers had no wish to alter the ways of the world, because of course they had no grasp of the fairly modern idea that the ways of the world might be altered. Yet what is one to make of a ballad like "The Broomfield Hill" (Child 43), in which male and female take on the qualities of combatants, and in which "true love" is so ironic? There are many fascinating variants of this ballad, some of which were recorded as early as the twelfth century; but the knight's attitude toward his "lady" strikes us as crudely contemporary.

A tryst is set among the broom, and a wager as well: the knight will give his "bonnie lass" five hundred pounds and ten if she meets him on the hill and is able to return to her home still a maiden. She accepts his offer with spirit:

> "I'll tak' your wager, bonnie lad,
> Five hunder pound and ten,
> That I'll gang tae the tap o' the hill
> And come back a maid again."

She goes out "greetin" (weeping) but comes in "lauchin," for she knows, evidently, of a way to triumph over her lover. When she meets him on Broomfield Hill—he is to be there at noon, she at one—she finds him asleep, and instead of waking him she circles him nine times, and then kisses his mouth; and leaves. (In other versions of the ballad the maiden uses a magic charm or herb to induce sleep in the knight; by walking around him nine times she exerts some sort of similar power over him.) When the knight wakes, he is angry with his friends for having allowed him to sleep through the tryst, and he says rather astonishingly—since, after all, the young woman is his "true love":

> "If ye'd hae waukened me frae my sleep,
> O' her I'd ha' taen my will,
> Though she'd hae deed the very next day,
> I would hae gotten my fill."

The ballad ends with a matter-of-fact summation that has, suddenly, nothing at all to do with love, but everything to do with the woman's triumph over the man, and the symbolic exchange of money. What strikes us as significant in "The Broomfield Hill," precisely because it is so horrifying—that a man would "get his fill" of a woman, even if she died as a consequence of his brutality—is brushed aside, as if so commonplace, so natural, that it does not require emphasis. The woman's triumphant laughter must be bitter; yet we hear only:

> So the wager's laid and the wager's paid,
> Five hunder pound and ten,
> 'Twas a' for her body's safety
> And the wager she did win.

But the appeal of the finest of the ballads—the often-anthologized "Sir Patrick Spens" and "The Twa Corbies," for instance—goes beyond the appeal of narrative, and even that of such disturbing ballads as "The Broomfield Hill" and "The Prickly Bush." The most beautiful ballads are not only stories but poems, and at times they demonstrate an insight—intelligent, moral, ironic—that appears to be artistically self-conscious. (Impossible to gauge, however, how many generations of anonymous singers passed before this consciousness was expressed in the ballad.) It is this "doubling back" upon the narrative situation, the explicit or implied comment on the story itself, that gives to the superior ballads their unquestionable value as art. And it is interesting indeed that this refinement of consciousness seems to question a tragic interpretation of human life.

"Tragic" elements in the ballads have more or less been taken for granted. The ballad scholar Francis Gummere located the ultimate appeal of the ballads in their portrayal of man's acceptance of fate: "Tragedy, not pessimism, is their last word. Their deepest value is that they revive to some extent the impressions which . . . communal poetry could make, by means now impossible for any poet to command."[2] It seems to me, however, that the "tragic" aura of the ballads is doubtful. Certainly they are filled with disasters, and with small sad domestic catastrophes; but the underlying consciousness of most of the ballads with which I am familiar, and particularly the most beautiful of the ballads, is not really sympathetic with a tragic view of life. The presence of "tragic situations" and an attitude that denies tragedy might seem contradictory; but an examination of a number of ballads that appear to possess tragic values will illustrate the distinction between the two—the simple presence of tragic incident in the narrative or fifth act, and the final ironical view of life of the lyric or chorus.

The "recognition" scene of tragedy is a very obvious device found in the ballads. The small group of ballads that concentrate on the illicit relationship between sister and brother often turns upon the disclosure of identity, which leads to death and despair. "Babylon" (Child 14) tells of a "banished man" who confronts three ladies, each of whom refuses him; he kills two of them, but is then told by the third that their brother will revenge them. When he learns that they are his own sisters he kills himself. "The Bonny Hind" (Child 50) tells of the discovery of an incestuous relationship after the girl and the man have become lovers; the girl commits suicide, the man returns to his father's house after burying her, and mourns for his "bonny hyn."

"Child Maurice" (Child 83) involves a husband's murder of a young man who has sent his wife tokens of love. The enraged husband meets Child Maurice in a "silver wood," fights with him, and beheads him:

> And he pricked itt on his swords poynt,
> Went singing there beside,
> And he rode till he came to *that* ladye ffaire,
> Whereas this ladye lyed.
>
> And sayes, Dost thou know Child Maurice head,
> If *that* thou dost itt see?
> And lapp itt soft, and kisse itt offt,
> Ffor thou louedst him better than mee.

> But when shee looked on Child Maurice head,
> She neuer spake words but three:
> "I neuer beare no child but one,
> And you haue slaine him trulye."

The husband then bitterly repents his deed. The situation, turning as it does upon a recognition of identity coming, of course, too late, could be called tragic. But an interesting variant of the ballad, Version B, suggests a far different consciousness of what the story means. The husband simply says to his wife, after she has revealed the identity of Child Maurice,

> "O wae be to thee, Lady Margaret," he sayd,
> "An ill death may you die;
> For if you had told me he was your son,
> He should neer have been slain by me."

The introduction of a very simple, common-sense attitude jars with the pretensions of "tragedy." The effect of the concluding lines is to undercut and negate the "tragic" situation entirely.

Situations in which the victim is passive can be called only loosely tragic. "Glasgerion" (Child 67) turns upon an incident that results in the deaths of three persons, two of them "nobility," but of which none rises to a level commensurate with tragedy. The situation of "Fair Mary of Wallington" (Child 91) is pathetic, in a way horrible, but there is no sense of a choice made which deliberately calls down fate upon the protagonist. The ballad begins with a swift contrast of moods:

> "When we were silly sisters seven,
> sisters were so fair
> Five of us were brave knights' wives,
> and died in childbirth lair."

The narrator of the first stanza apparently turns into "Fair Mary," who vows she will never marry; but when a knight seeks her for marriage she seems to accept at once the conditions of this marriage—the apparently inevitable death—without any real concern for the man involved:

> "If here's been the knight, mother,
> asking good will of me,
> Within three quarters of a year
> you may come bury me."

She is correct: but there is no indication of a conscious choice of love, a willingness to pay the price of death for this love; there is only the fatalistic passivity that these lines suggest. This same acceptance of death, with no consciousness of a choice of fates, is symptomatic of nearly all the ballads that involve death.

"Robin Hood's Death" (Child 120B), set entirely in the "fifth act" framework, occurs not as a result of heroic struggle against the sheriff's men, as one might expect, or as a result of Robin Hood's insisting upon attending Mass in spite of danger, but rather in a slow, passive, almost uneventful way. He says to Little John:

> ". . . I am not able to shoot one shot more,
> My broad arrows will not flee;
> But I have a cousin lives down below,
> Please God, she will bleed me."

Robin Hood dies from loss of blood and gives instructions to Little John concerning his death:

> "Let me have length and breadth enough,
> With a green sod under my head;
> That they may say, when I am dead,
> Here lies bold Robin Hood."

A more romantically heroic death is faced by "Johnie Armstrong" (Child 169), like Robin Hood an outlaw, who is killed in a fight with the king's men. Johnie has been deceived by the king, who has promised to pardon him:

> But Ionne looke'd over his left shoulder,
> Good Lord, what a grevious look looked hee!
> Saying, "Asking grace of a graceless face—
> Why, there is none for you nor me."

After he is wounded he says:

> "Fight on, my merry men all,
> And see that none of you be taine;
> For I will stand by and bleed but awhile,
> And then will I come and fight againe."

This spark of individuality, of a vital personality defying fate and at the same time accepting it, which is necessary to elevate character, is found in "Mary Hamilton" (Child 173), the story of a personal maid of the queen—from one of the highest families in Scotland—who "gangs wi bairn/To the hichest Stewart of a'." She is later condemned to death for murdering her infant child. She emerges as one of the few distinct personalities in all of the ballads. Told she must go to "Edinbro" to be sentenced, and to put on her black robe, she says to the queen herself:

> "I winna put on my robes o black,
> Nor yet my robes o brown;
> But I'll put on my robes o white,
> To shine through Edinbro town."

Going up the Canongate she "laughed loud laughters three," but coming down a "tear blinded her ee." Seeing ladies weeping for her,

"Ye need nae weep for me," she says
"Ye need nae weep for me;
For had I not slain mine own sweet babe,
This death I wadna dee."

Her statement of remorse is put in these words:

"Last nicht there was four Maries,
The nicht there'll be but three;
There was Marie Seton, and Marie Beton,
And Marie Carmichael, and me."

This combination of vitality and acceptance of fate is, however, not characteristic of the ballad heroes or heroines, whether they are of the primarily narrative or primarily lyric ballads. The tragic version of life seems to be neither that of the protagonists nor the unknown authors, though the incidents central to the works may seem tragic, or to possess a tragic potentiality.

"The Wife of Usher's Well" (Child 79) presents a situation "tragic" from only the mother's point of view; and this is why she mourns, and pronounces her curse:

"I wish the wind may never cease,
Nor flashes in the flood,
Till my three sons come hame to me,
In earthly flesh and blood."

The dead sons do return, but must leave again at daybreak. Version C introduces Christ to answer the woman's prayer by allowing her to receive her sons for a night; the sons then return to their tomb, and

"Go back, go back!" sweet Jesus replied,
"Go back, go back!" says he;
"For thou hast nine days to repent
For the wickedness that thou has done."

The woman's mourning and desire to see her sons again are taken to be "wicked"—an extremely unromantic moral judgment—apparently because they are against nature. That the woman is completely alone in her grief is evident from the extreme detachment of the poem—she is even called "carline" at one point—which not only objectively reports the cruelty of the world touching upon the woman, but sets up an implicit ironical antithesis between what the woman wants and what she can have, between her personal curse upon the universe and the ultimate impersonality of this universe. The "tragic" situation of the mother grieving for her dead sons loses its tragic edge by the focusing of the narrative not upon the individual so much as on the ironical tension between the wished-for and the real. Not only is the wife of Usher's Well denied tragic importance as an individual, but her situation itself is really denied tragic potential both through the detachment of the ballad and the explicit judgment which Christ makes in Variant C.

The well-known "The Three Ravens" (Child 26) also excludes a tragic consciousness of life, though for different reasons. The ballad is essentially a ballad of love, but one in which love is not equated with death in the way characteristic of a large number of sentimental and rather decadent love ballads; it presents instead a love which death cannot destroy, though this love seems to recognize the situation for what it is. There is no implied reunion after death, or a fulfillment of love through death, not even the perfunctory graveyard symbolism of entwining rose and brier, or birch and brier. The structure of "The Three Ravens" exemplifies the ideal ballad form: there is the framework of the ravens' conversation, whose words constitute the poem, and whose observant comments constitute the "story." All details save what happens in the present are omitted; the circumstances of the knight's death are not described; there is not even a description of him or of his lady, a use of the convenient, mechanical terms that often accompany knight and lady. The drama is given at its most elementary level, nearly a pantomime; it is empty of irony, empty of any real despair or undue mourning, though its climate is certainly not Christian —it is, if anything, pagan, though to call it so would be to misrepresent it seriously. "The Three Ravens," whose simplicity is not to be mistaken for simple-mindedness, is as fine as any of the ballads which do not attempt a refinement of perception or ironic consciousness found in the two or three outstanding works. It is most interesting in this particular discussion for its pure utilization of the "chorus," its translation of action into perception and comment, so that the "story" becomes flattened, compressed, given to us through dialogue, and the fifth act—indeed, any action at all—is realized only through the detached observations of the chorus of ravens. The transition from narrative to lyric has been made, here, but its great artistic potentiality in ballad art is achieved more completely by those ballads which include a certain backward-looking comment upon the story itself, a more self-conscious use of the chorus.

The surface similarity of "The Three Ravens" and its famous counterpart, "The Twa Corbies" (Child 26), suggests a relationship between the two that is deeper than that which does finally exist. It is a mistake to consider "The Twa Corbies" simply a cynical version of the earlier ballad, a reversal of the theme of true love;[3] it includes an objective presentation of an obvious "false" love, but its meaning goes far beyond this. Its meaning, at bottom, goes beyond all human concerns; and this is the terrible irony of the poem.

"The Twa Corbies" is shorter by half than "The Three Ravens," consisting of only five stanzas, and it has none of its visualized action or moralizing concern. It begins with the false or mechanical "I," the narrator who soon disintegrates:

As I was walking all alane,
I heard twa corbies making a mane;
The tane unto the t'other say,
"Where sall we gang and dine to-day?"

The complexity of the short poem is created through its ironic juxtaposition of two worlds: that of the human and that of the bestial. These worlds are overlapped—the corbies not only speak, but use such a formal word as "dine" to refer to their scavenger feeding. The world of human considerations, of a human contrivance of social and moral systems, is presented in terms of its negatives, for here the knight's hound, his hawk, and his lady have deserted him; they have not preserved the faithful love the three ravens had earlier observed. The knight has, in a double sense, departed from the civilized world, for his death is unknown: ". . . naebody kens that he lies there,/But his hawk, his hound, and lady fair." Here the term "lady fair" takes on an additional note of irony simply because it is so standard a phrasing, and implies perhaps a universal collaboration in betrayal among all "ladies fair." The almost mechanical summation of the knight's desertion by conventional symbols of faith and love is jarringly culminated by a conclusion in terms of a practical evaluation of these three betrayals:

His hound is to the hunting gane,
His hawk to fecht the wild-fowl hame,
His lady's ta'en another mate,
So we may mak our dinner sweet.

The impact of the faithlessness of hound, hawk, and lady, and the apparent treachery of the lady, are powerful in themselves, but they are overshadowed by the deeper significance of the poem, which presents the human in a situation devoid of all social, moral, or religious aids, devoid of all civilized values. The acts of love which soften the horror of death in

"The Three Ravens" have vanished in "The Twa Corbies." The knight as a social being has become, here, no more than an incidental constituent of the natural process of consumption and decay. The grandeur of knighthood is reduced to organic material. The corbie says,

> "Ye'll sit on his white hause-bane,
> And I'll pike out his bonny blue een;
> Wi ae lock o his gowden hair
> We'll theek our nest when it grows bare."

All is seen from the scavengers' point of view. The peculiar irony of the poem grows out of such lines as "We'll theek our nest when it grows bare," which suggests a concern on the part of the corbies for their own future creaturely comfort, and a permanence of relation denied the knight. The final vision of the ballad is one which outdistances all considerations of personal tragedy. It is not only the vanity of human values but the forlorn emptiness of the natural universe itself that echoes in the lines, stressed by the slowness of the long "a" vowels:

> Mony a one for him makes mane,
> But nane sall ken where he is gane;
> O'er his white banes, when they are bare,
> The wind sall blaw for evermair.

Narrative has here become lyric. The corbies are the "chorus" of the drama, but the drama itself—the human tragedy—is already finished. Human tragedy is a matter of particulars, and will vanish; but the chorus of nature, of the natural universe, will not vanish, and its slow, sure, deadly vision constitutes an ironic consciousness of the way the world is.

That the ballad has here become lyric is evident from the fact that, as a story of a particular event—like, for instance, "The Hunting of the Cheviot" Child 162)—"The Twa Corbies" is meaningless. The symbolic structuring, clearer here than in most of the finer ballads, suggests definitely both a conscious artistry and a concern for metaphoric meaning. A brief flash of something of this penetration into an evaluation of order and lawlessness is found in Variant E of "Sheath and Knife" (Child 16). The lines are those of stanza nine:

> The hawk had nae lure, and the horse had nae master,
> And the faithless hounds thro the woods ran faster.

"Sheath and Knife," however, in all its variants, does not approach the mastery of "The Twa Corbies." What is central is not the fact of death itself, but the indifferent response to individual death in a naturalistic universe. The emphasis is upon what we would call the unromantic and the untragic, a certain blurring of tragic action into broader significance, expressed by the detached chorus of the corbies—the old men or women of the village.

A synthesis of the "fifth act" of revelation and the detachment of the chorus is achieved in the famous "Edward" (Child 13B). This ballad has as its literal center the fact of death, but as its metaphorical center the illumination, by degrees, of a bitter irony. The obvious use of incremental repetition, which leads structurally to the revelation of both patricide and a woman's treachery, follows a pattern, like "Lord Randal" (Child 12), of an ordering or revelations in terms of their relative importance. Thus the climax is built up through increasingly significant answers, and the structural technique with its deliberate slowness of movement gives the work its emotional impact:

> "Why dois your brand sae drap wi bluid,
> Edward, Edward,
> Why dois your brand sae drap wi bluid,
> And why sae sad gang yee O?"

"O I hae killed my hauke sae guid,
And I had nae mair bot hee O."

The mother is not satisfied with the answer and questions him further. He answers that he has killed his "reid-roan steid," but this also does not satisfy her. He admits finally that he has killed his "fadir deir." The mechanical formality of the work, however, will allow for no pause after this revelation. The movement is dance-like and formal, and must continue; and the mother immediately asks after his penance—

". . . whatten penance will ye drie for that?
My deir son, now tell me O."
"Ile set my feit in yonder boat,
Mither, mither,
Ile set my feit in yonder boat,
And Ile fare ovir the sea O."

—and then after his legacies. His answers reveal his increasingly violent state of mind. He will let his towers stand "tul they doun fa," he will leave to his wife and children "the warldis room, late them beg thrae life." The climax of the series of disclosures is this reply to his mother:

"And what wul ye leive to your ain mither deir?
My deir son, now tell me O."
"The curse of hell frae me sall ye beir,
Mither, mither,
The curse of hell frae me sall ye beir,
Sic counseils ye gave to me O."

Again, the repetitive nature of the technique, the refrain of "Mither, mither," adds to the now revealed irony. Like "Lord Randal," the dialogue framework will allow for no episodic action, and the formalized diction, the emphatic rhythm of the words, contrast with the enormity of the crime committed and the final revelation of the collaboration of the mother in this crime. The gradual building-up in two parts —to the disclosure of the father's death, and then to the disclosure of the mother's "counseils"—with its flawless gradation, certainly argues a very sophisticated artistry. Beyond simply the "narrative" dialogue there is a conscious concern for the evoking of a certain emotional response in the audience, which ballad scholars have not, in general, recognized. Even if through oral transmission details of the ballad are changed—the substitution of hound for hawk, for instance— the essential structural grasp of the ballad cannot change, or it would lose its meaning as an artistic unit. Again, the instance of death here does not suggest a tragic evaluation of life; the fact of death is flatly stated, a matter of the past, and it is the flash of irony at the end that makes the ballad so successful. In it the actors of the fifth act and the chorus— the formalized, almost detached revealers of what has happened—are more or less fused.

"The Twa Corbies" and "Edward," two masterpieces of ballad art, are matched in the English and Scottish ballads only by "Sir Patrick Spens" (Child 58). The most famous version, A, is one of the shortest, and it gains from its compactness and lack of transitional and expositional material. Indeed, what is explicitly told in the ballad is of less *literal* importance than what is omitted—we know the color of the king's wine, and that the Scottish ladies have gold combs in their hair, but we know nothing at all about what should be the central event of the ballad as narrative, the wrecking of the ship. The argument that the ballad art contains only the most essential details of an action is, then, clearly not applicable to the very finest ballads, if generally applicable to any. Actual detail is absent; we have instead symbolic detail, details which have meanings beyond themselves and which move toward, of course, a greater significance than literal or factual detail would suggest. The narrative content of the work is obviously suppressed; it is not a story that the ballad tells, surprisingly enough, but a revelation, finally, of a certain vision of life, a lyric expression of a view of death and of the vanity of the world.

The irony of the ballad is expressed through a series of contrasts. There is, first, the king who "sits in Dumferling toune,/Drinking the blude-reid wine," who appears to be honoring Sir Patrick by asking him to make this important voyage.[4] The ironic juxtaposition of the king in town and Sir Patrick, later, at sea, is matched by the king's drinking of the "blood-red wine": an obvious symbol of the king's utilization of Sir Patrick (and of any of his subjects) on a very casual and predatory level. The irony is enhanced by the king's apparent lack of malice, his lack, indeed, of any consideration of Sir Patrick as a fellow human being. The first great irony is that between arbitrary ruler and subject.

Sir Patrick, reading the king's letter, reacts with mixed feelings:

> The first line that Sir Patrick red,
> A loud lauch lauched he;
> The next line that Sir Patrick red,
> The teir blinded his ee.

The impending voyage is both ridiculous and tragic: Sir Patrick, as a man of apparently magnanimous or at least worldly consciousness, must recognize and appreciate both the ridiculousness and the tragedy. The time of the year is dangerous, and one of his crewmen extends Sir Patrick's judgment by speculation upon the weather:

> "Late late yestreen I saw the new moone,
> Wi the auld moone in hir arme,
> And I feir, I feir, my deir master,
> That we will cum to harme"

The irony of the situation is that the men must go to sea with a full awareness of their fate; and this awareness is so certain, so much taken for fact, that the next stanza of the poem treats only the lightest and apparently least essential details about the Scots nobles:

> O our Scots nobles wer richt laith
> To weet their cork-heild schoone;
> Bot lang owre a' the play wer playd,
> Thair hats they swam aboone.

There is no depiction of the sinking of the ship, only the unmistakable meaning of the final line. And the movement of the ballad is decidedly away from the singularity of this catastrophe:

> O Lang, lang may their ladies sit,
> Wi thair fans into their hand,
> Or eir they se Sir Patrick Spence
> Cum sailing to the land.
>
> O lang, lang may the ladies stand,
> Wi thair gold kems in their hair,
> Waiting for thair ain der lords,
> For they'll se thame na mair.
>
> Haf owre, haf owre to Aberdour,
> It's fiftie fadom deip,
> And thair lies guid Sir Patrick Spence,
> Wi the Scots lords at his feit.

The tragic potential is here suppressed and dissolved; if there is a "tragedy" it is not an uncommon one, and it is undercut at nearly every turn by a latent and very powerful irony. Thus the men who are not ordinary seamen, but Scots nobles, are visualized in terms of their apparel and their concern for this apparel. Their own vain concern is taken, ironically, by the narrator as a legitimate estimation of their own worth, and they and their ladies emerge as ghostly figures with costly apparel and no significant lasting worth. The men are loath to wet their cork-heeled shoes, but their hats swim ashore: between these two statements the whole disaster occurs, but it is seen in only these apparently light observations concerning shoes and hats, as if these were really the most essential details to characterize the men. And there is the paradox of the hats which could, after all, swim ashore, though the men themselves could not. So with the ladies who wait with their fans in hand and their gold combs in their hair while their men die at sea. The final irony of the poem is that which groups the Scots nobles at the feet of Spens. The concern for rank and order suggested by the cork-heeled shoes and the combs of gold is parodied by this mock observation of rank fifty fathoms under the sea.

So it is that the sparseness of "essential" detail about the catastrophe is finally unimportant. It is not a narrative the ballad tells, but rather a statement about the human predicament. The series of ironical contrasts undercuts all pretensions to tragedy or even to an understanding of the pathos of what has happend. Many of the English and Scottish popular ballads do begin, as Gray said, in the fifth act; but a number of them begin when even this fifth act is over, when the tragic actors have left the stage and only the chorus remains to give universal and objective meaning to what has happened. The lyric expression of the chorus, whether the unknown narrator of "Sir Patrick Spens" or the corbies in "The Two Corbies," is in a sense directly *opposed* to the tragic action that has occurred or is occurring, though irreparably related to it. The distinction is simply one of point of view. Where to the tragic actors the situation is particularized and tragic, to the "chorus" it is related to the abstract and symbolic, the timeless and cyclical reenactment of the way the world is. The individual may perish, but Nature endures: to experience its non-human, indeed inhuman, presence is to share in its divinity. The vision of the "folk"—the Unconscious—transcends history and is always contemporary with us. *We* are the folk, immortal.

Notes

1. The English scholar Francis James Child published his massive textual compilation, *The English and Scottish Popular Ballads,* from 1882 to 1898; the number in parentheses after each ballad title is that assigned to the ballad by Child.

2. Francis Barton Gummere, *The Popular Ballad* (New York, 1907), pp. 340–41.

3. See Earl Daniels, *The Art of Reading Poetry* (New York, 1941), p. 133.

4. According to Professor Child, this ballad may or may not be historical. The ballad versions fall into two classes—the first giving little or no "historical" information, the second giving additional details. Thus in the second group the destination of Spens's ship is Norway; the object of the voyage (not told in Variant G) is to bring home the king of Norway's daughter (or the Scottish king's daughter), or to take the Scottish king's daughter to Norway, where she is to be queen.

Lawrence's Götterdämmerung: *The Apocalyptic Vision of* Women in Love

> *And was he fated to pass away in this knowledge, this one process of frost-knowledge, death by perfect cold? Was he a messenger, an omen of the universal dissolution into whiteness and snow?*
>
> *Birkin thinking of Gerald,* Women in Love

In a little-known story of Lawrence's called "The Christening" an elderly wreck of a man contemplates his illegitimate grandchild and attempts to lead his embarrassed and impatient household in a prayer in "the special language of fatherhood." No one listens, no one wishes to hear. He is rambling, incoherent, bullying even in his confession and self-abnegation, yet his prayer is an extraordinary one: he implores God to shield the newborn child from the conceit of family life, from the burden of being a *son* with a specific *father.* It was his own interference with his children, his imposition of his personal will, that damaged them as human beings; and he prays that his grandson will be spared this violation of the spirit. Half-senile he insists upon his prayer though his grown-up children are present and resentful:

> "Lord, what father has a man but Thee? Lord, when a man says he is a father, he is wrong from the first word. For Thou art the Father, Lord. Lord, take away from us

> the conceit that our children are ours. . . . For I have stood between Thee and my children; I've had *my* way with them, Lord; I've stood between Thee and my children; I've cut 'em off from Thee because they were mine. And they've grown twisted, because of me. . . . Lord, if it hadn't been for me, they might ha' been trees in the sunshine. Let me own it, Lord, I've done 'em mischief. It would ha' been better if they'd never known no father."

Between the individual and the cosmos there falls the deathly shadow of the ego: the disheveled old man utters a truth central to Lawrence's work. Where the human will is active there is always injury to the spirit, always a perversion, a "twisting"; that human beings are compelled not only to assert their greedy claims upon others but to manipulate their own lives in accord with an absolute that has little to do with their deeper yearnings constitutes our tragedy. Is it a tragedy of the modern era; is it inevitably bound up with the rise of industry and mechanization? Lawrence would say that it is, for the "material interests" of which Conrad spoke so ironically are all that remain of spiritual hopes; God being dead, God being unmasked as a fraud, nothing so suits man's ambition as a transvaluing of values, the reinterpretation of religious experience in gross, obscene terms. Here is Gerald Crich, one of Lawrence's most deeply realized and sympathetic characters, surely an alter ego of his—

> In his travels, and in his accompanying readings, he had come to the conclusion that the essential secret of life was harmony. . . . And he proceeded to put his philosophy into practice by forcing order into the established world, translating the mystic word harmony into the practical word organisation.[1]

Harmony becomes *organization.* And Gerald dedicates himself to work, to feverish, totally absorbing work, inspired with an almost religious exaltation in his fight with matter.

The world is split in two: on one side matter (the mines, the miners), on the other side his own isolated will. He wants to create on earth a perfect machine, "an activity of pure order, pure mechanical repetition"; a man of the twentieth century with no nostalgia for the superannuated ideals of Christianity or democracy, he wishes to found his eternity, his infinity, in the machine. So inchoate and mysterious is the imaginative world Lawrence creates for *Women in Love* that we find no difficulty in reading Gerald Crich as an allegorical figure in certain chapters and as a quite human, even fluid personality in others. As Gudrun's frenzied lover, as Birkin's elusive beloved, he seems a substantially different person from the Gerald Crich who is a ruthless god of the machine; yet as his cultural role demands extinction (for Lawrence had little doubt that civilization was breaking down rapidly, and Gerald is the very personification of a "civilized" man), so does his private emotional life, his confusion of the individual will with that of the cosmos, demand death—death by perfect cold. He is Lawrence's only tragic figure, a remarkable creation in a remarkable novel, and though it is a commonplace to say that Birkin represents Lawrence, it seems equally likely that Gerald Crich represents Lawrence—in his deepest, most aggrieved, most nihilistic soul.

Women in Love is an inadequate title. The novel concerns itself with far more than simply *women* in love; far more than simply women *in love*. Two violent love affairs are the plot's focus, but the drama of the novel has clearly to do with every sort of emotion, and with every sort of spiritual inanition. Gerald and Birkin and Ursula and Gudrun are immense figures, monstrous creations out of legend, out of mythology; they are unable to alter their fates, like tragic heroes and heroines of old. The mark of Cain has been on Gerald since early childhood, when he accidentally killed his brother; and Gudrun is named for a heroine out of Germanic legend who slew her first husband. The pace of the novel is often frenetic. Time is running out, history is coming to an end, the Apocalypse is at hand. *Dies Irae* and *The Latter Days* (as well as *The Sisters* and *The Wedding Ring*) were titles Lawrence considered for the novel, and though both are too explicit, too shrill, they are more suggestive of the chiliastic mood of the work (which even surprised Lawrence when he read it through after completion in November of 1916: it struck him as "end-of-the-world" and as "purely destructive, not like *The Rainbow*, destructive-consummating").[2]

Women in Love is a strangely ceremonial, even ritualistic work. In very simple terms it celebrates love and marriage as the only possible salvation for twentieth-century man and dramatizes the fate of those who resist the abandonment of the ego demanded by love: a sacrificial rite, an ancient necessity. Yet those who "come through"—Birkin and Ursula—are hardly harmonious; the novel ends with their arguing about Birkin's thwarted desire for an "eternal union with a man," and one is given to feel that the shadow of the dead man will fall across their marriage. And though the structure of the novel is ceremonial, its texture is rich, lush, fanciful, and, since each chapter is organized around a dominant image, rather self-consciously symbolic or imagistic; action is subordinate to theme. The perversity of the novel is such that its great subject of mankind's tragically split nature is demonstrated in the art-work itself, which is sometimes a fairly conventional novel with a forward-moving plot, sometimes a gorgeous, even outrageous prose poem on the order of the work Aloysius Bertrand and Charles Baudelaire were doing in the previous century. Birkin is sometimes a prophetic figure, and sometimes merely garrulous and silly; Ursula is sometimes a mesmerizing archetypal female, at other times shrill and possessive and dismayingly obtuse. In one of Lawrence's most powerful love scenes Gerald Crich comes by night to Gudrun's bedroom after his father's death and is profoundly revitalized by her physical love, but Gudrun cannot help looking upon him with a devastating cynicism, noting his ridiculous trousers and braces and boots, and she is filled with nausea of him despite her fascination. Gudrun herself takes on in Gerald's obsessive imagination certain of the more destructive qualities of the Magna Mater or the devouring female, and she attains an almost mythic power over him; but when we last see her she has become shallow and cheaply ironic, merely a vulgar young woman. It is a measure of Lawrence's genius that every part of his immensely ambitious novel works (with the possible exception of the strained chapter "In The Pompadour") and that the proliferating images coalesce into fairly stable leitmotifs: water, moon, darkness, light, the organic and the sterile.

Our own era is one in which prophetic eschatological art has as great a significance as it did in 1916; Lawrence's despairing conviction that civilization was in the latter days is one shared by a number of our most serious writers, even if there is little belief in the Apocalypse in its classical sense. The notion of antichrist is an archaic one, a sentiment that posits unqualified belief in Christ; and the ushering in of a violent new era, a millennium, necessitates faith in the transcendental properties of the world, or the universe, which contrast sharply with scientific speculations about the fate we are likely to share. Even in his most despairing moments Lawrence remained curiously "religious." It is a tragedy that Western civilization may be doomed, that a man like Gerald Crich must be destroyed, and yet—does it really matter? Lawrence through Birkin debates the paradox endlessly. He cannot come to any conclusion. Gerald is beloved, yet Gerald is deathly. Gerald is a brilliant young man, yet he is a murderer, he is suicidal, he is rotten at the core. It is a possibility that Birkin's passionate love for him is as foully motivated as Gudrun's and would do no good for either of them. *Can* human beings alter their fates? Though his pessimism would seem to undercut and even negate his art, Lawrence is explicit in this novel about his feelings for mankind; the vituperation expressed is perhaps unequaled in serious literature. Surely it is at the very heart of the work, in Birkin's strident ranting voice:

> "I detest what I am, outwardly. I loathe myself as a human being. Humanity is a huge aggregate lie, and a huge lie is less than a small truth. Humanity is less, far less than the individual, because the individual may sometimes be capable of truth, and humanity is a tree of lies. . . .
>
> " . . . I abhor humanity, I wish it was swept away. It could go, and there would be no *absolute* loss, if every human being perished to-morrow."

But Ursula also perceives in her lover a contradictory desire to "save" this doomed world, and characteristically judges this desire a weakness, and insidious form of prostitution. Birkin's perverse attachment to the world he hates is not admirable in Ursula's eyes, for Ursula is no ordinary woman but a fiercely intolerant creature who detests all forms of insincerity. She is Birkin's conscience, in a sense; his foil, his gadfly; a taunting form of himself. Yet later, immediately after Birkin declares that he loves her, she is rather disturbed by the starkly nihilistic vision he sets before her; and indeed it strikes us as more tragic than that of Shakespeare:

> "We always consider the silver river of life, rolling on and quickening all the world to a brightness, on and on to heaven, flowing into a bright eternal sea, a heaven of angels thronging. But the other is our real reality . . . that dark river of dissolution. You see it rolls in us just as the other rolls—the black river of corruption. And our flowers are of this—our sea-born Aphrodite, all our white phosphorescent flowers of sensuous perfection, all our reality, nowadays."

Aphrodite herself is symptomatic of the death-process, born in what Lawrence calls the "first spasm of universal dissolution." The process cannot be halted. It is beyond the individual, beyond choice. It ends in a universal nothing, a new cycle in which humanity will play no role. The prospect is a chilling one and yet—*does* it really matter? Humanity in the aggregate is contemptible, and many people (like Diana Crich) are better off dead since their living has somehow gone wrong. No, Birkin thinks, it can't *really* matter. His mood shifts, he is no longer frustrated and despairing, he is stoical, almost mystical, like one who has given up all hope. For he has said earlier to Gerald, after their talk of the death of God and the possible necessity of the salvation through love, that reality lies outside the human sphere:

> "Well, if mankind is destroyed, if our race is destroyed like Sodom, and there is this beautiful evening with the luminous land and trees, I am satisfied. That which informs it all is there, and can never be lost. After all, what is mankind but just one expression of the incomprehensible. And if mankind passes away, it will only mean that this particular expression is completed and done. . . . Humanity doesn't embody the utterance of the incomprehensible any more. Humanity is a dead letter. There will be a new embodiment, in a new way. Let humanity disappear as quick as possible."

Lawrence's shifts in mood and conviction are passionate, even unsettling. One feels that he writes to discover what he thinks, what is thinking in him, on an unconscious level. Love is an ecstatic experience. Or is it, perhaps, a delusion? Erotic love is a way of salvation—or is it a distraction, a burden? Is it something to be gone through in order that one's deepest self may be stirred to life? Or is it a very simple, utterly natural emotion . . . ? (In *Sons and Lovers* Paul Morel is impatient with Miriam's near-hysterical exaggeration of ordinary emotions; he resents her intensity, her penchant for mythologizing, and finds solace in Clara's far less complex attitude toward sexual love.) Lawrence does not really know, regardless of his dogmatic remarks about "mind-consciousness" and "blood-consciousness." He cannot *know;* he must continually strive to know, and accept continual frustration.[3]

Tragedy for Lawrence arises out of the fatal split between the demands of the ego and those of the larger, less personal consciousness: we are crippled by the shadow of the finite personality as it falls across our souls, as the children of the old man in "The Christening" are crippled by his *particular* fatherliness. If at one point in history—during the great civilization of the Etruscans, for instance—there was a unity of being, a mythic harmony between man and his community and nature, it is lost to us now; the blighted landscapes in Beldover through which Lawrence's people walk give evidence that humanity is no longer evolving but devolving, degenerating. ("It is like a country in an underworld," says Gudrun, repulsed but fascinated. "The people are all ghouls, and everything is ghostly. Everything is a ghoulish replica of the real world . . . all soiled, everything sordid. It's like being mad, Urusla.") One England blots out another England, as Lawrence observes in *Lady Chatterley's Lover* some years later.

In Lawrence's work one is struck repeatedly by the total absence of concern for community. In the novels after *Sons and Lovers* his most fully developed and self-contained characters express an indifference toward their neighbors that is almost aristocratic. Both Anna and Will Brangwen of *The Rainbow* are oblivious to the world outside their household: the nation does not exist to them; there is no war in South Africa; they are in a "private retreat" that has no nationality. Even as a child Ursula is proudly contemptuous of her classmates, knowing herself set apart from them and, as a Brangwen, superior. She is fated to reject her unimaginative lover Skrebensky who has subordinated his individuality to the nation and who would gladly give up his life to it. ("I belong to the nation," he says solemnly, "and must do my duty by the nation.") Some years later she and Gudrun express a loathing for their parents' home that is astonishing, and even the less passionate Alvina Houghton of *The Lost Girl* surrenders to outbursts of mad, hilarious jeering, so frustrated is she by the limitations of her father's household and of the mining town of Woodhouse in general. (She is a "lost" girl only in terms of England. Though her life in a primitive mountain village in Italy is not a very comfortable one, it is nevertheless superior to her former, virginal life back in provincial England.)

Lawrence might have dramatized the tragedy of his people's rootlessness, especially as it compels them to attempt desperate and often quixotic relationships as a surrogate for social and political involvement (as in *The Plumed Serpent* and *Kangaroo*); but of course he could not give life to convictions he did not feel. The human instinct for something larger than an intense, intimate bond, the instinct for community, is entirely absent in Lawrence, and this absence helps to account for the wildness of his characters' emotions. (Their passionate narrowness is especially evident when contrasted with the tolerance of a character like Leopold Bloom of *Ulysses*. Leopold thinks wistfully of his wife, but he thinks also of innumerable other people, men and women both, the living and the dead; he is a man of the city who is stirred by the myriad trivial excitements of Dublin—an adventurer writ small, but not contemptible in Joyce's eyes. His obsessions are comically perverse, his stratagems pathetic. Acceptance by Simon Dedalus and his friends would mean a great deal to poor Bloom, but of course this acceptance will be withheld; he yearns for community but is denied it.)

For the sake of argument Gudrun challenges Ursula's

conviction that one can achieve a new space to be in, apart from the old: "But don't you think you'll *want* the old connection with the world—father and the rest of us, and all that it means, England and the world of thought—don't you think you'll *need* that, really to make a world?" But Ursula speaks for Lawrence in denying all inevitable social and familial connections. "One has a sort of other self, that belongs to a new planet, not to this," she says. The disagreement marks the sisters' break with each other; after this heated discussion they are no longer friends. Gudrun mocks the lovers with her false enthusiasm and deeply insults Ursula. "Go and find your new world, dear. After all, the happiest voyage is the quest of Rupert's Blessed Isles."

Lawrence's utopian plans for Rananim aside, it seems obvious that he could not have been truly interested in establishing a community of any permanence, for such a community would have necessitated a connection between one generation and the next. It would have demanded that faith in a reality beyond the individual and the individual's impulses which is absent in Lawrence—not undeveloped so much as simply absent, undiscovered. For this reason alone he seems to us distinctly un-English in any traditional sense. Fielding and Thackeray and Trollope and Dickens and Eliot and Hardy and Bennett belong to another world, another consciousness entirely. (Lawrence's kinship with Pater and Wilde, his predilection for the intensity of the moment, may have stimulated him to a vigorous glorification of Nietzschean instinct and will to power as a means of resisting aestheticism: for there is a languid cynicism about Birkin not unlike that of Wilde's prematurely weary heroes.)

Halfway around the world, in Australia, Richard Somers discovers that he misses England, for it isn't freedom but mere *vacancy* he finds in this new, disturbingly beautiful world: the absence of civilization, of culture, of inner meaning; the absence of spirit.[4] But so long as Lawrence is in England he evokes the idea of his nation only to do battle with it, to refute it, to be nauseated by it. The upper classes are sterile and worthless, the working classes are stunted aborigines who stare after the Brangwen sisters in the street. Halliday and his London friends are self-consciously decadent—"the most pettifogging calculating Bohemia that ever reckoned its pennies." Only in the mythical structure of a fabulist work like *The Escaped Cock* can Lawrence imagine a harmonious relationship between male and female, yet even here in this Mediterranean setting the individual cannot tolerate other people, nor they him: "the little life of jealousy and property" resumes its sway and forces the man who died to flee. There is, however, no possibility of a tragic awareness in these terms; it is not tragic that the individual is compelled to break with his nation and his community because any unit larger than the individual is tainted and suspect, caught in the downward process of corruption.[5] The community almost by definition is degraded. About this everyone is in agreement—Clifford Chatterley as well as Mellors, Hermione as well as Ursula and Gudrun. Community in the old sense is based on property and possessions and must be rejected, and all human relationships not founded upon an immediate emotional rapport must be broken. "The old ideals are dead as nails—nothing there," Birkin says early in *Women in Love*. "It seems to me there remains only this perfect union with a woman—sort of ultimate marriage—and there isn't anything else." Gerald, however, finds it difficult to agree. Making one's life up out of a woman, one woman only, woman only seems to him impossible, just as the forging of an intense love-connection with another man—which in Lawrence's cosmology would have saved his life—is impossible.

"I only feel what I feel," Gerald says.

The core of our human tragedy has very little to do with society, then, and everything to do with the individual: with the curious self-destructive condition of the human spirit. Having rejected the theological dogma of original sin, Lawrence develops a rather similar psychological dogma to account for the diabolic split within the individual between the dictates of "mind-consciousness" and the impulses of "blood-consciousness." In his essay on Nathaniel Hawthorne in *Studies in Classic American Literature*, he interprets *The Scarlet Letter* as an allegory, a typically American allegory, of the consequences of the violent antagonism between the two ways of being. His explicitness is helpful in terms of *Women in Love*, where a rich verbal texture masks a tragically simple paradox. The cross itself is the symbol of mankind's self-division, as it is the symbol, the final haunting image, in Gerald Crich's life. (Fleeing into the snow, exhausted and broken after his ignoble attempt to strangle Gudrun, Gerald comes upon a half-buried crucifix at the top of a pole. He fears that someone is going to murder him. In terror he realizes "This was the moment when the death was uplifted, and there was no escape. Lord Jesus, was it then bound to be—Lord Jesus! He could feel the blow descending, he knew he was murdered.")

Christ's agony on the cross symbolizes our human agony at having acquired, or having been poisoned by, the "sin" of knowledge and self-consciousness. In the Hawthorne essay Lawrence says:

> Nowadays men do hate the idea of dualism. It's no good, dual we are. The cross. If we accept the symbol, then, virtually we accept the fact. We are divided against ourselves.
>
> For instance, the blood *hates* being KNOWN by the

> mind. It feels itself destroyed when it is KNOWN. Hence the profound instinct of privacy.
>
> And on the other hand, the mind and the spiritual consciousness of man simply *hates* the dark potency of blood-acts: hates the genuine dark sensual orgasms, which do, for the time being, actually obliterate the mind and the spiritual consciousness, plunge them in a suffocating flood of darkness.
>
> You can't get away from this.
>
> Blood-consciousness overwhelms, obliterates, and annuls mind-consciousness.
>
> Mind-consciousness extinguishes blood-consciousness, and consumes the blood.
>
> We are all of us conscious in both ways. And the two ways are antagonistic in us.
>
> They will always remain so.
>
> That is our cross.

It is obvious that Lawrence identifies with the instinct toward formal allegory and subterfuge in American literature. He understands Hawthorne, Melville, and Poe from the inside; it is himself he speaks of when he says of Poe that he adventured into the vaults and cellars and horrible underground passages of the human soul, desperate to experience the "prismatic ecstasy" of heightened consciousness and of love. And Poe knew himself to be doomed, necessarily—as Lawrence so frequently thought himself (and his race). Indeed, Poe is far closer to Lawrence than Hawthorne or Melville:

> He died wanting more love, and love killed him. A ghastly disease, love. Poe telling us of his disease: trying even to make his disease fair and attractive. Even succeeding. Which is the inevitable falseness, duplicity of art, American art in particular.

The inevitable duplicity of art: an eccentric statement from the man who says, elsewhere (in an essay on Walt Whitman), and the essential function of art is moral. "Not aesthetic, not decorative, not pasttime and recreation. But moral." Yet it is possible to see that the artist too suffers from a tragic self-division, that he is forced to dramatize the radically new shifting over of consciousness primarily in covert, even occult and deathly terms: wanting to write a novel of consummate health and triumph whose controlling symbol is the rainbow, writing in fact a despairing, floridly tragic and rather mad work that resembles poetry and music (Wagnerian music) far more than it resembles the clearly "moral" bright book of life that is the novel, Lawrence finds himself surprised and disturbed by the apocalyptic nature of this greatest effort, as if he had imagined he had written something quite different. The rhythm of Lawrence's writing is that of the American works he analyzes so irreverently and so brilliantly, a "disintegrating and sloughing of the old consciousness" and "the forming of a new consciousness underneath." Such apocalyptic books must be written because old things need to die, because the "old white psyche has to be gradually broken down before anything else can come to pass" (in the essay on Poe). Such art must be violent, it must be outlandish and diabolic at its core because it is revolutionary in the truest sense of the word. It is subversive, even traitorous; but though it seeks to overturn empires, its primary concerns are prophetic, even religious. As Lawrence says in the poem "Nemesis" (from *Pansies),* "If we do not rapidly open all the doors of consciousness/and freshen the putrid little space in which we are cribbed/the sky-blue walls of our unventilated heaven/will be bright red with blood." In any case the true artist does not determine the direction of his art; he surrenders his ego so that his deeper self may be heard. There is no freedom except in compliance with the spirit within, what Lawrence calls the Holy Ghost.

The suppressed Prologue to *Women in Love* sets forth the terms of Birkin's torment with dramatic economy.[6]

"Mind-consciousness" and "blood-consciousness" are not mere abstractions, pseudo-philosophical notions, but bitterly existential ways of perceiving and of being. When Birkin and Gerald Crich first meet they experience a subtle bond between each other, a "sudden connection" that is intensified during a mountain-climbing trip in the Tyrol. In the isolation of the rocks and snow they and their companion attain a rare sort of intimacy that is to be denied and consciously rejected when they descend again into their unusual lives. (The parallel with Gerald's death in the snow is obvious; by suppressing the Prologue and beginning with the chapter we have, "Sisters," in which Ursula and Gudrun discuss marriage and the home and the mining town and venture out to watch the wedding, Lawrence sacrified a great deal. "Sisters" is an entirely satisfactory opening, brilliant in its own lavish way; but the Prologue with its shrill, tender, almost crazed language is far more moving.)

Preliminary to the action of *Women in Love,* and unaccountable in terms of *The Rainbow,* which centers so exclusively upon Ursula, is the passionate and undeclared relationship between Birkin and Gerald, and the tortured split between Birkin's spiritual and "sisterly" love for Hermione and his "passion of desire" for Gerald. Birkin is sickened by his obsession with Gerald; he is repulsed by his overwrought, exclusively mental relationship with Hermione (which is, incidentally, very close to the relationship of sheer nerves Lawrence discusses in his essay on Poe: the obscene love that is the "intensest nervous vibration of unison" without any erotic consummation). That Birkin's dilemma is emblematic of society's confusion in general is made clear, and convincing, by his immersion in educational theory. What is education except the gradual and deliberate building up of consciousness, unit by unit? Each unit of consciousness is the "living unit of that great social, religious, philosophic idea towards which mankind, like an organism seeking its final form, in laboriously growing," but the tragic paradox is that there *is* no great unifying idea at the present time; there is simply aimless, futile activity. For we are in the autumn of civilization, and decay, as such, cannot be acknowledged. As Birkin suffers in his awareness of his own deceitful, frustrated life, he tries to forget himself in work; but he cannot escape a sense of the futility to all attempts at "social constructiveness." The tone of the Prologue is dark indeed, and one hears Lawrence's undisguised despair in every line:

> How to get away from this process of reduction, how escape this phosphorescent passage into the tomb, which was universal though unacknowledged, this was the unconscious problem which tortured Birkin day and night. He came to Hermione, and found with her the pure, translucent regions of death itself, of ecstasy. In the world the autumn itself was setting in. What should a man add himself on to?—to science, to social reform, to aestheticism, to sensationalism? The whole world's constructive activity was a fiction, a lie, to hide the great process of decomposition, which had set in. What then to adhere to?

He attempts a physical relationship with Hermione which is a cruel failure, humiliating to them both. He goes in desperation to prostitutes. Like Paul Morel he suffers a familiar split between the "spiritual" woman and the "physical" woman, but his deeper anxiety lies in his unacknowledged passion for Gerald Crich. Surely homoerotic yearning has never been so vividly and so sympathetically presented as it is in Lawrence's Prologue, where Birkin's intelligent complexity, his half-serious desire to rid himself of his soul in order to escape his predicament, and his fear of madness and dissolution as a consequence of his lovelessness give him a tragic depth comparable to Hamlet's. He *wants* to love women, just as he wants to believe in the world's constructive activity; but how can a man create his own feelings? Birkin knows that he cannot: he can only suppress them by an act of sheer will. In danger of going mad or of dying—of possibly killing himself—Birkin continues his deathly relationship with Hermione, keeping his homoerotic feelings to himself and even, in a sense, secret from himself. With keen insight Lawrence analyzes Birkin's own analysis of the situation. "He knew what he felt, but he always kept the knowledge at bay. His a priori were: 'I *should* not feel like this,' and 'It is the ultimate mark of my own deficiency, that I feel like this.' Therefore, though he admitted everything, he never really faced the question. He never accepted the desire, and received it as part of himself. He always tried to keep it expelled from him." Not only does Birkin attempt to dissociate himself from an impulse that *is* himself, he attempts to deny the femaleness in his own nature by objectifying (and degrading) it in his treatment of Hermione and of the "slightly bestial" prostitutes. It maddens him that he should feel sexual attraction for the male physique while for the female he is capable of feeling only a kind of fondness, a sacred love, as if for a sister. "The women he seemed to be kin to, he looked for the soul in them." By the age of thirty he is sickly and dissolute, attached to Hermione in a loveless, sadistic relationship, terrified of breaking with her for fear of falling into the abyss. Yet the break is imminent, inevitable—so the action of *Women in Love* begins.

A tragedy, then, of an informal nature, experimental in its gropings toward a resolution of the central crisis: how to integrate the male and female principles, how to integrate the organic and the "civilized," the relentlessly progressive condition of the modern world. It is not enough to be a child of nature, to cling to one's ignorance as if it were a form of blessedness; one cannot deny the reality of the external world, its gradual transformation from the Old England into the New, into an enthusiastic acceptance of the individual as an instrument in the great machine of society.

When Hermione goes into her rhapsody about spontaneity and the instincts, echoing Birkin in saying that the mind is death, he contradicts her brutally by claiming that the problem is not that people have too much mind, but too little. As for Hermione herself, she is merely making words because knowledge means everything to her: "Even your animalism, you want it in your head. You don't want to *be* an animal, you want to observe your own animal functions, to get a mental thrill out of them. . . . What is it but the worst and last form of intellectualism, this love of yours for passion and the animal instincts?" But it is really himself he is attacking: Hermione is a ghastly form of himself he would like to destroy, a parody of a woman, a sister of his soul.

Women in Love must have originally been imagined as Birkin's tragedy rather than Gerald's, for though Gerald feels an attraction for Birkin, he is not so obsessed with it as Birkin is; in the Prologue he is characterized as rather less intelligent, less shrewd, than he turns out to be in subsequent chapters. Ursula's role in saving Birkin from dissolution is, then, far greater than she can know. Not only must she arouse and satisfy his spiritual yearnings, she must answer to his physical desire as well: she must, in a sense, take on the active, masculine role in their relationship. (Significantly, it is Ursula who presses them into an erotic relationship after the death of Diana Crich and her young man. It is she who embraces Birkin tightly, wanting to show him that she is no shallow prude, and though he whimpers to himself, "Not this, not this," he nevertheless succumbs to desire for her and they become lovers. Had Ursula not sensed the need to force Birkin into a physical relationship, it is possible their love would have become as spiritualized, and consequently as poisoned, as Birkin's and Hermione's.) Ursula's role in saving Birkin from destruction is comparable to Sonia's fairly magical redemption of Raskolnikov in *Crime and Punishment*, just as Gerald's suicide is comparable to Svidrigaylov's when both men are denied salvation through women by whom they are obsessed. Though the feminine principle is not sufficient to guarantee eternal happiness, it is nevertheless the way through which salvation is attained: sex is an initiation in Lawrence, a necessary and even ritualistic *event* in the process of psychic wholeness. Where in more traditional tragedy—Shakespeare's *King Lear* comes immediately to mind—it is the feminine, irrational, "dark and vicious" elements that must be resisted, since they disturb the status quo, the patriarchal cosmos, in Lawrence it is precisely the darkness, the passion, the mind-obliterating, terrible, and even vicious experience of erotic love that is necessary for salvation. The individual is split and wars futilely against himself, civilization is split and must fall into chaos if male and female principles are opposed. Lawrence's is the sounder psychology, but it does not follow that his world view is more optimistic, for to recognize a truth does not inevitably bring with it the moral strength to realize that truth in one's life.

Birkin's desire for an eternal union with another man is thwarted in *Women in Love*, and his failure leads indirectly to Gerald's death. At least this is Birkin's conviction. "He should have loved me," he says to Ursula and she, frightened, replies without sympathy, "What difference would it have made!" It is only in a symbolic dimension that the men are lovers; consciously, in the daylight world, they are never anything more than friends. In the chapter "Gladiatorial" the men wrestle together in order to stir Gerald from his boredom, and they seem to "drive their white flesh deeper and deeper against each other, as if they would break into a oneness." The effort is such that both men lose consciousness and Birkin falls over Gerald, involuntarily. When their minds are gone their opposition to each other is gone and they can become united—but only temporarily, only until Birkin regains his consciousness and moves away. At the novel's conclusion Birkin is "happily" married, yet incomplete. He will be a reasonably content and normal man, a husband to the passionate Ursula, yet unfulfilled; and one cannot quite believe that his frustrated love for Gerald will not surface in another form. His failure is not merely his own but civilization's as well: male and female are inexorably opposed, the integration of the two halves of the human soul is an impossibility in our time.[7]

❧

Hence the cruel frost-knowledge of *Women in Love,* the death by perfect cold Lawrence has delineated. Long before Gerald's actual death in the mountains Birkin speculates on him as a strange white wonderful demon from the north, fated like his civilization to pass away into universal dissolution, the day of "creative life" being finished. In *Apocalypse* Lawrence speaks of the long slow death of the human being in our time, the victory of repressive and mechanical forces over the organic, the pagan. The mystery religions of antiquity have been destroyed by the systematic, dissecting principle; the artist is driven as a consequence to think in deliberately mythical, archaic, chiliastic terms. How to express the inexpressible? Those poems in *Pansies* that address themselves to the problem—poems like "Wellsian Futures," "Dead People," "Ego-Bound," "Climb Down, O Lordly Mind," "Peace and War"—are rhetorical and strident and rather flat; it is in images that Lawrence *thinks* most clearly. He is too brilliant an artist not to breathe life even into those characters who are in opposition to his own principles. In a statement that resembles Yeats's (that the occult spirits of *A Vision* came to bring him images for his poetry) Lawrence indicates a surprising indifference to the very concept of the Apocalypse itself: "We do not care, vitally, about theories of the Apocalypse. . . . What we care about is the release of the imagination. . . . What does the Apocalypse matter, unless in so far as it gives us imaginative release into another vital world?"[8]

This jaunty attitude is qualified by the images that are called forth by the imagination, however: the wolfishness of Gerald and his mother; the ghoulishness of the Beldover miners; the African totems (one has a face that is void and terrible in its mindlessness; the other has a long, elegant body with a tiny head, a face crushed small like a beetle's); Hermione striking her lover with a paperweight of lapis lazuli and fairly swooning with ecstasy; Gerald digging his spurs into his mare's sides, into wounds that are already bleeding; the drowned Diana Crich with her arms still wrapped tightly about the neck of her young man; the demonic energy of Winifred's rabbit, and Gudrun's slashed, bleeding arm which seems to tear across Gerald's brain; the uncanny, terrifying soullessness of Innsbruck; the stunted figure of the artist Loerke; the final vision of Gerald as the frozen carcass of a dead male. These are fearful images, and what has Lawrence to set against them but the embrace of a man and a woman, a visionary transfiguration of the individual by love?—and even the experience of love, of passion and unity, is seen as ephemeral.

Birkin sees Gerald and Gudrun as flowers of dissolution, locked in the death-process; he cannot help but see Gerald as Cain, who killed his brother. Though in one way *Women in Love* is a naturalistic work populated with realistic characters and set in altogether probable environments, in another way it is inflexible and even rather austerely classical: Gerald is Cain from the very first and his fate is settled. Birkin considers his friend's accidental killing of his brother and wonders if it is proper to think in terms of *accident* at all. Has everything that happens a universal significance? Ultimately he does not believe that there is anything accidental in life: "it all hung together, in the deepest sense."

(And it follows that no one is murdered accidentally: ". . . a man who is murderable is a man who in a profound if hidden lust desires to be murdered.") Gerald plainly chooses his murderer in Gudrun, and it is in the curious, misshapen form of Loerke that certain of Gerald's inclinations are given their ultimate realization. Gerald's glorification of the machine and of himself as a god of the machine is parodied by Loerke's inhuman willfulness: Gudrun sees him as the rock-bottom of all life. Unfeeling, stoic, he cares about nothing except his work, he makes not the slightest attempt to be at at one with anything, he exists a "pure, unconnected will" in a stunted body. His very being excites Gerald to disgusted fury because he is finally all that Gerald has imagined for himself—the subordination of all spontaneity, the triumph of "harmony" in industrial organization.

Of the bizarre nightmare images stirred in Lawrence's imagination by the idea of the Apocalypse, Loerke is perhaps the most powerful. He is at once very human, and quite inhuman. He is reasonable, even rather charming, and at the same time deathly—a "mud-child," a creature of the underworld. His name suggests that of Loki, the Norse god of discord and mischief, the very principle of dissolution. A repulsive and fascinating character, he is described by Lawrence as a gnome, a bat, a rabbit, a troll, a chatterer, a magpie, a maker of disturbing jokes, with the blank look of inorganic misery behind his buffoonery. That he is an artist, and a homosexual as well, cannot be an accident. He is in Lawrence's imagination the diabolic alter ego who rises up to mock all that Lawrence takes to be sacred. Hence his uncanny power, his parodistic talent: he accepts the hypothesis that industry has replaced religion and he accepts his role as artist in terms of industry, without sentimental qualms. Art should interpret industry; the artist fulfills himself in acquiescence to the machine. Is there nothing apart from work, mechanical work?—Gudrun asks. And he says without hesitation, "Nothing but work!"

Loerke disgusts Birkin and Gerald precisely because he embodies certain of their own traits. He is marvelously self-sufficient; he wishes to ingratiate himself with no one; he is an artist who completely understands and controls his art; he excites the admiration of the beautiful Gudrun, and even Ursula is interested in him for a while. Most painful, perhaps, is his homosexuality. He is not divided against himself, not at all tortured by remorse or conscience. In the Prologue to the novel Birkin half-wishes he might rid himself of his soul, and Loerke is presented as a creature without a soul, one of the "little people" who finds his mate in a human being. It is interesting to note that the rat-like qualities in Loerke are those that have attracted Birkin in other men: Birkin has felt an extraordinary desire to come close to and to know and "as it were to eat" a certain type of Cornish man with dark, fine, stiff hair and dark eyes like holes in his head or like the eyes of a rat (see the Prologue); and he has felt the queer, subterranean, repulsive beauty of a young man with an indomitable manner "like a quick, vital rat" (see the chapter "A Chair"). The Nietzschean quality of Loerke's haughtiness and his loathing of other people, particularly women, remind us of the aristocratic contempt expressed by the middle-aged foreigner whom Tom Brangwen admires so much in the first chapter of *The Rainbow:* the man has a queer monkeyish face that is in its way almost beautiful, he is sardonic, dry-skinned, coldly intelligent, mockingly courteous to the women in his company (one of whom has made love with Tom previously), a creature who strangely rouses Tom's blood and who, in the form of Anna Lensky, will be his mate. There is no doubt but that Lawrence, a very different physical type, and temperamentally quite opposed to the cold, life-denying principle these men embody, was nevertheless powerfully attracted by them. There is an irresistible *life* to Loerke that makes us feel the strength of his nihilistic charm.

Surely not accidental is the fact that Loerke is an artist. He expresses a view of art that all artists share, to some extent, despite their protestations to the contrary. It is Flaubert speaking in Loerke, declaring art supreme and the artist's life of little consequence; when Loerke claims that his statuette of a girl on a horse is no more than an artistic composition, a certain form without relation to anything outside itself, he is echoing Flaubert's contention that there is no such thing as a subject, there is only style. ("What seems beautiful to me, what I should like to write," Flaubert said, in a remark now famous, "is a book about nothing, a book dependent on nothing external. . . .") Loerke angers Ursula by declaring that his art pictures nothing, "absolutely nothing," there is no connection between his art and the everyday world, they are two different and distinct planes of existence, and she must not confuse them. In his disdainful proclamation of an art that refers only to itself, he speaks for the aesthetes of the nineteenth century against whom Lawrence had to define himself as a creator of vital, moral, life-enhancing art. Though Lawrence shared certain of their beliefs—that bourgeois civilization was bankrupt, that the mass of human beings was hopelessly ignorant and contemptible—he did not want to align himself with their extreme rejection of "ordinary" life and of nature itself. (Too unbridled a revulsion against the world would lead one to the sinister self-indulgent fantasies of certain of the decadent poets and artists—the bizarre creations of Oscar Wilde and Huysmans and Baudelaire, and of Gustave Moreau and Odilon Redon and Jan Toorop among others.) Loerke's almost supernatural presence drives Ursula and Birkin away, and brings to the surface the destructive elements in the love of Gudrun and Gerald. He is an artist of decay: his effect upon Gudrun is like that of a subtle poison.

"Life doesn't *really* matter," Gudrun says. "It is one's art which is central."[9]

Symbolically, then, Gerald witnesses the destruction of his love, or of a part of his own soul, by those beliefs that had been a kind of religion to him in his operating of the mines. Lawrence himself plays with certain of his worst fears by giving them over to Loerke and Gudrun, who toy with them, inventing for their amusement a mocking dream of the destruction of the world: humanity invents a perfect explosive that blows up the world, perhaps; or the climate shifts and the world goes cold and snow falls everywhere and "only white creatures, polar-bears, white foxes, and men like awful white snow-birds, persisted in ice cruelty." It is Lawrence's nightmare, the Apocalypse without resurrection, without meaning; a vision as bleak and as tragically unsentimental as Shakespeare's.

Only in parable, in myth, can tragedy be transcended. In that beautiful novella *The Escaped Cock*, written while Lawrence was dying, the Christian and the pagan mate, the male and the female come together in a perfect union, and the process of dissolution is halted. The man who had died awakes in his tomb, sickened and despairing, knowing himself moral, not the Son of God but no more than a son of man—and in this realization is his hope, his true salvation. He is resurrected to the flesh of his own body; through the warm, healing flesh of the priestess of Isis he is healed of his fraudulent divinity. "Father!" he cries in his rapture, "Why did you hide this from me?"

Poetic, Biblical in its rhythms, *The Escaped Cock* is an extraordinary work in that it dramatizes Lawrence's own sense of resurrection from near death (he had come close to dying several times) and that it repudiates his passion for changing the world. The man who had died realizes that his teaching is finished and that it had been a mistake to interfere in the souls of others; he knows now that his reach ends in his fingertips. His love for mankind had been no more than a form of egotism, a madness that would devour multitudes while leaving his own being untouched and virginal. What is crucified in him is his passion for "saving" others. Lawrence has explored the near dissolution of the personality in earlier works—in Ursula's illness near the end of *The Rainbow,* and in her reaction to Birkin's love-making in *Women in Love;* and in Connie Chatterley's deepening sense of nothingness before her meeting with Mellors—but never with such powerful economy as in *The Escaped Cock.* The man who had died wakes slowly and reluctantly to life, overcome with a sense of nausea, dreading consciousness but compelled to return to it and to his fulfillment as a human being. The passage back to life is a terrible one; his injured body is repulsive to him, as is the memory of his suffering. The analogy between the colorful cock and the gradually healing flesh of the man who had died is unabashedly direct and even rather witty. In this idyllic Mediterranean world a cock and a man are kin, all of nature is related, the dead Osiris is resurrected in the dead Christ, and the phenomenal world is revealed as the transcendental world, the world of eternity. Simply to live in a body, to live as a mortal human being—this is enough, and this is everything. Only a man who had come close to dying himself and who had despaired of his efforts to transform the human world could have written a passage like this, in awed celebration of the wonders of the existential world:

> The man who had died looked nakedly onto life, and saw a vast resoluteness everywhere flinging itself up in stormy or subtle wave-crests, foam-tips emerging out of the blue invisible, a black-and-orange cock, or the green flame tongues out of the extremes of the fig-tree. They came forth, these things and creatures of spring, glowing with desire and with assertion. . . . The man who had died looked on the great swing into existence of things that had not died, but he saw no longer their tremulous desire to exist and to be. He heard instead their ringing, defiant challenge to all other things existing. . . . And always, the man who had died saw not the bird alone, but the short, sharp wave of life of which the bird was the crest. He watched the queer, beaky motion of the creature.
> . . .
>
> And the destiny of life seemed more fierce and compulsive to him even than the destiny of death.

The man who had died asks himself this final question: *From what, and to what, could this infinite whirl be saved?*

The mystic certitude of *The Escaped Cock,* like the serenity of "The Ship of Death" and "Bavarian Gentians," belongs to a consciousness that has transcended the dualism of tragedy. The split has not been healed, it has simply been transcended; nearing death, Lawrence turns instinctively to the allegorical mode, the most primitive and the most sophisticated of all visionary expressions. *Women in Love* is, by contrast, irresolute and contradictory; it offers only the finite, tentative "resurrection" of marriage between two very incomplete people. Like Connie Chatterley and her lover Mellors, the surviving couple of *Women in Love* must fashion their lives in a distinctly unmythic, unidyllic landscape, their fates to be bound up closely with that of their civilization. How are we to escape history?—defy the death-process of our culture? With difficulty. In sorrow. So long as we live, even strengthened as we are by the "mystic conjunction," the "ultimate unison" between men and women, our lives are tempered by the ungovernable contingencies of the world that is no metaphor, but our only home.

Notes

1. All quotations from *Women in Love* are taken from the Modern Library edition.

2. *Collected Letters,* ed. Harry T. Moore (New York, 1962), pp. 482 and 519.

3. As Lawrence says in an essay about the writer's relationship to his own work: "Morality in the novel is the trembling instability of the balance. When the novelist puts his thumb on the scale, to pull down the balance to his own predilection, that is immorality. . . . And of all the art forms, the novel most of all demands the trembling and oscillating of the balance," *Phoenix: The Posthumous Papers of D. H. Lawrence* (London, 1936), p. 529.

4. Richard Somers is fascinated and disturbed by Australia, into which he has projected the struggle of his own soul. The bush has frightened him with its emptiness and stillness; he cannot penetrate its secret. At one point it seems to him that a presence of some sort lurks in the wilderness, an actual spirit of the place that terrifies him. As for the social and political conditions of Australia—what is more hopelessly uninteresting than accomplished liberty? (See *Kangaroo,* London, 1968, p. 33.)

5. That Lawrence might have dealt with the tragic implications of the individual's failure to find a home for himself in his own nation is indicated by remarks he makes elsewhere, for instance in the introductory essay, "The Spirit of Place," to *Studies in Classic American Literature:* "Men are free when they are in a living homeland, not when they are straying and breaking away. Men are free when they are obeying some deep, inward voice of religious belief. Obeying from within. Men are free when they belong to a living, organic, *believing* community, active in fulfilling some unfulfilled, perhaps unrealized purpose." The *Studies* were written between 1917 and 1923.

6. The Prologue is available in *Phoenix II* (New York, 1968) and in a recently published anthology, *The Other Persuasion,* ed. Seymour Kleinberg (New York, 1977).

7. It is interesting to note Lawrence's intense dislike of the very idea of homosexuality in women. Miss Inger of *The Rainbow* is revealed as a poisonous, corrupt woman who makes an ideal mate for Ursula's cynical uncle Tom Brangwen. Ursula had loved them both but when she realizes that they are in the "service of the machine," she is repulsed by them. "Their marshy, bitter-sweet corruption came sick and unwholesome in her nostrils. . . . She would leave them both forever, leave forever their strange, soft, half-corrupt element" (*The Rainbow,* London, 1971, p. 351). In *Lady Chatterley's Lover* Mellors begins to rant about women he has known who have disappointed him sexually, and the quality of his rage—which must be, in part, Lawrence's—is rather alarming. He goes through a brief catalogue of unacceptable women, then says, "It's astonishing how Lesbian women are, consciously or unconsciously. Seems to me they're nearly all Lesbian." In the presence of such a woman, Mellors tells Connie, he fairly howls in his soul, "wanting to kill her" (*Lady Chatterley's Lover,* New York, 1962, p. 190).

8. *Phoenix,* pp. 293–94.

9. Gudrun is an artist of considerable talent herself, one who works in miniatures, as if wishing to see the world "through the wrong end of the opera glasses." It is significant that she expresses a passionate wish to have been born a man, and that she feels an unaccountable lust for deep brutality against Gerald, whom in another sense she loves. Far more interesting a character than her sister Ursula, Gudrun is fatally locked into her own willful instinct for making herself the measure of all things: her vision is anthropomorphic and solipsistic, finally inhuman. We know from certain of Lawrence's poems, particularly "New Heaven and Earth," that the "maniacal horror" of such solipsism was his own. He seems to have been driven nearly to suicide, or to a nervous breakdown, by the terrifying conviction that nothing existed beyond his own consciousness. Unlike Lawrence, who sickened of being the measure of all things, Gudrun rejoices in her cruel talent for reducing everyone and everything—robins as well as people—to size. Her love affair with Gerald is really a contest of wills; in her soul she is a man, a rival. Like one of the seductive chimeras or vampires in decadent art—in the paintings of Munch and in the writings of Strindberg–Gudrun sees her lover as an "unutterable enemy," whom she wishes to kiss and stroke and embrace until she

has him "all in her hands, till she [has] strained him into her knowledge. Ah, if she could have precious *knowledge* of him, she would be filled . . ." (379). At the novel's end she has become so dissociated from her own feelings and so nauseated by life that she seems to be on the brink of insanity. It strikes her that she has never really lived, only worked, she is in fact a kind of clock, her face is like a clock's face, a twelve-hour clock dial—an image that fills her with terror, yet pleases her strangely.

Jocoserious Joyce

"Everything speaks in its own way."
Ulysses

Ulysses is certainly the greatest novel in the English language, and one might argue for its being the greatest single work of art in our tradition. How significant, then, and how teasing, that this masterwork should be a comedy and that its creator should have explicitly valued the comic "vision" over the tragic—how disturbing to our predilection for order that, with an homage paid to classical antiquity so meticulous that it is surely a burlesque, Joyce's exhibitionististicicity is never so serious as when it is most outrageously comic. Joyce might have been addressing his readers when he wrote to Nora in 1909: "Now . . . I want you to read over and over all I have written to you. Some of it is ugly, obscene, and bestial, some of it is pure and holy and spiritual: all of it is myself."

After a few reluctant, abortive, and surely humiliating attempts on Joyce's part to accommodate himself to others' wishes—in regard to the initial publication of *Dubliners* most of all—Joyce was thrown back upon his own idiosyncratic genius, and was freed to invent works of a kind never before attempted in literature. Fortunately for him he had, as his brother Stanislaus ironically noted, "that inflexibility firmly rooted in failure." *Silence, exile, cunning—and failure:* would a Joyce whose first novel was *Stephen Hero* and whose *Dubliners* had been sufficiently diminished (by its young author) to qualify for publication in Dublin have advanced to *Ulysses?* Had *Stephen Hero* been published, what of *A Portrait of the Artist as a Young Man,* whose telegraphic, lyric, contemplative style evolves quite naturally into the early style of *Ulysses?* Such speculations are fruitless but irresistible, especially as they radiate out to touch upon all artists and all artistic endeavor. Early and continued success may be the destruction of all but the most ingenious or melancholy of writers: in any case, Joyce was never confronted with this problem. And he was further charged, as we know, by the zestful energies of revenge—a revenge all the more satisfying because it is never savage, but witty and graceful and so subtle as to nearly pass us by when we encounter it, as in this brief exchange between Buck Mulligan and Haines at tea in the "Wandering Rocks" episode—

> ——Ten years, he said, chewing and laughing. He is going to write something in ten years.
>
> ——Seems a long way off, Haines said, thoughtfully lifting his spoon. Still, I shouldn't wonder if he did after all.

Conceived, perhaps, as a work enlivened by Swiftian satire, *Ulysses* must have evolved as the years passed and passages were written and rewritten (and always expanded) into a symphony of voices, a profane comedy of Dublin and Dubliners that—the Citizen of "Cyclops" not overlooked—is at bottom deeply sympathetic with each voice. The righteous indignation of satire is very difficult for a novelist to sustain, especially if the novelist is embarked upon a lengthy project in which each "voice" makes some claim for its individuality. The precarious formality and asceticism of tragedy is also difficult to sustain or, in fact, to take seriously, when one has written and rewritten and contemplated and rewritten again those climactic passages that affirm the "tragic sense of life." With so much time to dream his paean to Dublin into being, with a certain relentless self-scrutiny that was no doubt cultivated in Joyce by his Jesuit teachers, Joyce could not have failed to see, as time passed, how greatly he had come to differ from the near-humorless Stephen of *Portrait,* how much more tolerant his own nature was; he is willing to share the lofty title of "artist" with his own Leopold Bloom (of whom it is said "There's a bit of the artist about old Bloom") and willing to risk mocking his own strenuous efforts to create a masterpiece, in such passages as this, from "Ithaca":

> What were habitually [Bloom's] final meditations?
>
> Of some one sole unique advertisement to cause passers to stop in wonder, a poster novelty, with all extraneous accretions excluded, reduced to its simplest and most efficient terms not exceeding the span of casual vision and congruous with the velocity of modern life.

And Molly's ostensibly far-fetched idea of being the inspiration for Stephen Dedalus's inevitable fame isn't really—if one is aware of Joyce's hopes for himself and Nora—so far-fetched after all, but marvelously witty:

> . . . they all write about some woman in their poetry well I suppose he wont find many like me . . . Im sure itll be grand if I can only get in with a handsome young poet at my age . . . Ill read and study all I can find or learn a bit off my heart if I knew who he likes so he wont think me stupid if he thinks all women are the same and I can teach him the other part Ill make him feel all over him till he half faints under me then hell write about me lover and mistress publicly too with our 2 photographs in all the papers when he becomes famous O but then what am I going to do about him though

Knowing what we do of Joyce's tendency to lift from other writers situations and plots and themes and even bits of language—not to mention immense structures like that of Homer's—it is significant that Bloom (who is in reality not a writer) should unwisely boast of "following a literary occupation" in the "Nighttown" sequence. It is not Bloom but his creator who is the "author-journalist" engaged in "just bringing out a collection of prize stories of which I am the inventor, something that is an entirely new departure." He is subsequently denounced by Mr. Philip Beaufoy, who accuses him not only of plagiarism but of not being a gentleman (another of Joyce's—and John Joyce's—fixed notions):

> BEAUFOY: (*Drawls.*) No, you aren't, not by a long shot if I know it. I don't see it, that's all. No born gentleman, no one with the most rudimentary promptings of a gentleman would stoop to such particularly loathsome conduct. One of those, my lord. A plagiarist. A soapy sneak masquerading as a literateur. It's perfectly obvious that with the most inherent baseness he has cribbed some of my bestselling books, really gorgeous stuff, a perfect gem, the love passages in which are beneath suspicion. . . . My literary agent Mr J. B. Pinker is in attendance.

Stephen has said to Lynch, in *Portrait,* that it is the artist's personality that dominates the work of art, though the artist must be, of course, refined out of existence so that his shadow does not crudely fall upon his creation. This personality "passes into the narration itself, flowing round and round the persons and the action like a vital sea" and—in *Ulysses* certainly—it seems to blend with its material, to become ever more protean and magical as the long day progresses. The "voice" of the narrator conditions and is in turn conditioned by the fictional voices he recalls or re-creates—many of them, of course, living people; friends of Joyce's and of his father's—and Joyce has said in a letter that "each adventure (every hour, every organ, every art being interconnected and interrelated in the structural scheme of the whole) should not only condition but even create its own technique": so that we have, in *Ulysses* as in no other work except *Finnegans Wake,* the alarming phenomenon of a novel continually "in progress."

Comedy is never gracious to fixed forms; the comic spirit is by its nature youthful and irreverent and rebellious. Critical emphasis that has been placed upon the Homeric structure in *Ulysses* is misleading, then, at least for the reader who is relatively unfamiliar with Joyce's way of going about things, for it thrusts into the foreground a kind of scaffolding, an ingenious apparatus, and asks us to admire it, when *Ulysses* might have evolved about any number of "myths" in the public domain. (Joyce did, in fact, think of using the "Faust" legend as a means of structuring his Dubliners' adventures—indifferent to the radical discrepancies between an "Odysseus" and a "Faust"; the novelist is, surely, the most ingeniously pragmatic of all artists.) The gigantic structure of the work must have been felt by Joyce—consciously or half-consciously—as a real necessity, a way of protecting its creator from the sort of playful, fussy multiplication of details that spring to life in "Nighttown" and would have overloaded and cracked his narrative if not restrained elsewhere. The rebellious impulse does, after all, require a fixed and even rather tyrannical structure against which to rebel; playfulness is only italicized against someone else's gravity. T. S. Eliot could not really have understood *Ulysses* or Joyce when he made his statement in *The Dial* (now famous, of course, as all Eliot's statements seem to have become—the erroneous no less than the helpful) that the use of myth as a way of "manipulating a continuous parallel between modernity and antiquity" was a method which other writers *must* pursue, following Joyce's example; one might argue that Joyce cared essentially little for whatever is meant by "modernity" and "antiquity," and that, in any case, the Joycean method is not a method at all but a process of personality, inimitable. (When consciously imitated, as in *The Waste Land,* the method seems to be a means of fouling the present by a continuous and highly romantic if not somewhat mad "parallel" with various pasts and various traditions. Any slander upon life as it is routinely lived is antithetical to Joyce's sensibility.) Eliot could respond to the intellectual qualities of *Ulysses,* and to its satirical passages, but he was not capable of responding to the heart of the work itself: a Mass or celebration of Dublin on June 16, 1904. One day being all days, one city being all cities, one "plot" doing as well as any plot, with characters, of course, continually walking through themselves (in Stephen's words) "meeting robbers, ghosts, giants, old men, young men, wives, widows, brothers-in-love," but always meeting themselves. For God—whether the hangman God of Christianity or the ingenious Artist-God—is "doubtless all in all in all of us." The highest and most spirited comedy is by necessity democratic—even anarchic. It celebrates life: the livingness of life, not its abstract qualities. Where Eliot saw the contemporary world as futile because disruptive of the past, Joyce, the realist-fantasist, the unparalleled mimic, gave life to these clamorous voices without passing judgment on them.

"I do not think that any writer has yet presented Dublin to the world," Joyce said in a letter of 1905 (accompanying the ill-fated manuscript of *Dubliners*). The Dublin of his short stories did not satisfy him, for its "paralysis," its "special odor of corruption" were only a part of the Dublin he remembered. Thus *Ulysses:* an immense sanctification of the city and its inhabitants, a dizzying kaleidoscope of images, a cacophony of voices, an unlikely harmony. The artist's wish to remember perfectly, to transcribe the "reality of experience," is heightened by his intense love for his own background and for the fast-fading voices of his father's world. The spirit of Dublin *is* the voice of *Ulysses.* All the narrators are aspects of this single voice: Dublin is important because it belongs to them, and it is through Dublin that the narrators achieve their own immortality. In *Ulysses* everything has its own way of speaking, its own music. People are not really distinct from the complex pattern of relationships that is the human world at any given moment nor are they distinct from the landscape itself, which will come more aggressively alive in *Finnegans Wake.*

Where tragedy demands a paring back of contingencies so that whatever fate overtakes the hero appears to be inevitable, comedy of the complex sort Joyce was attempting can appear to arise spontaneously out of accidents and misunderstandings—a good example being the "Throwaway" incident, when poor Bloom inadvertently gives Bantam Lyons a tip on the Gold Cup race and, hours later, finds himself an object of bitter resentment in Barney Kiernan's pub. Comic fate is—as Bloom thinks in bed—"more than inevitable, irreparable." It has more to do with local gossip than with the gods' designs. There is something uncanny about *Ulysses,* the presence of ghost-narrators who impede or occasionally—as in the deadly "Eumaeus" chapter—actually distort the narrative, erecting a barrier between the reader and the characters, of whom he has become rather fond as the long day unfolds. Had *Ulysses* ended with "Eumaeus" it would have been one of the most depressing works ever written, for in this chapter the narrator breaks down much of what the other narrators have created. Joyce, knowing himself a supreme stylist, is not afraid here to expose willfully his own methods by belaboring them. The overexplicitness of the chapter, quite apart from the chapter's subjects of illusion and lies and "types" (W. B. Murphy and his tall tales; Bloom and his presentation of Molly as a "Spanish type") risks undercutting all that has gone before. The brisk telegraphic "stream-of-consciousness" of earlier chapters is here slowed to a perverse, maddening tempo ("He began to remember that this had happened, or had been mentioned as having happened, before but it cost him no small effort before he remembered that he recognized . . .") and the novel's life-blood, its marvelously interrelated contingencies, is exposed and found to be commonplace after all ("Yet still, though his eyes were thick with sleep and sea air, life was full of a host of things and coincidences of a terrible nature and it was quite within the bounds of possibility that it was not an entire fabrication though at first blush there was not much inherent probability in all the spoof he got off his chest being strictly accurate gospel"). Is it possible that the anonymous narrators of *Ulysses* are meant to suggest supernatural presences, of the sort that are ubiquitous in Homer? If so, Athena in such chapters works a kind of counter-magic, undoing the splendors that have gone before. While tragedy must end at precisely the right moment, comedy can take us a little farther along; we are invited to observe the artist's labor and even to comment upon it. Joyce populates a world and then shows it to be inside his head, a stratagem like one of Bloom's.

Elsewhere, the narrators' voices are recognizable as distinct from Joyce's and there is no danger of our confusing them, nor is there any danger of their intruding upon the reader's sense of the novel's worth. The Gerty MacDowell sequence is too famous to require comment; Gerty is less experienced and less pragmatic than Molly, but it might be instructive to read her voice as a variation of the voice of the "sane full amoral fertilisable untrustworthy engaging limited prudent indifferent *Weib*." As Milly is a younger version of Molly (the similarity of their names suggesting the short-hand characterizations of comic opera), so Gerty may be a virginal type of Molly; and her romanticism, clotted and pathetic as it is, calls to mind the more effusive passages of *Giacomo Joyce*. The narrator of the chapter knows Bloom, however, whom he calls—as all the narrators do—*Mr.* Bloom (in contrast to their more familiar "Stephen"). The narrator of the *entr'acte* chapter, "Wandering Rocks," has anticipated Gerty MacDowell and her involvement with Bloom. In this chapter simultaneous events, though dispersed in space, intrude casually upon one another; it is as if the narrator himself were the wanderer, moving freely about Dublin and environs and seeing (without valuing that he sees) a dizzying profusion of interrelationships, some of which are not made clear for hundreds of pages. (We learn the identity of the "flushed young man" and the young woman with "wild nodding daisies" whom Father Conmee blesses much, much later, and the knowledge is inconsequential.) That the presiding spirit of "Nighttown" forces upon Bloom and Stephen images, phrases, and epiphanies not their own argues for its being a spirit of place rather than a means by which the "inner lives" of the protagonists are dramatized. For instance, Cissy Caffrey and Edy Boardman appear early in the episode, and they assuredly do not belong in "Nighttown" or in either Bloom's or Stephen's imaginations. Bloom's witty "Shoot him! Dog of a christian! So much for M'Intosh!" is really out of Stephen's imagination; the gnomic prayers of the Daughters of Erin which run lightly through the chapters of *Ulysses* cannot possibly spring out of Bloom; and the nonsense about the "second-best bed" relates back to Stephen's hypothesis concerning Shakespeare and his relationship with Ann Hathaway, and has nothing at all to do with Bloom. The vision of Rudy with which "Nighttown" concludes is generally held to be a serious one, Bloom's own image of his beloved son as he would be had he lived; yet it seems to me that the vision is a mocking one, a grotesque parody of Bloom's fatherhood. How can this "Rudy" be anything other than the narrator's stylized invention, with which he confronts the wonder-struck Bloom as if wishing to free him of his sterile obsession?—

> Against the dark wall a figure appears slowly, a fairy boy of eleven, a changeling, kidnapped, dressed in an Eton suit with glass shoes and a little bronze helmet, holding a book in his hand. He reads from right to left inaudibly, smiling, kissing the page. . . .
> Gazes unseeing into Bloom's eyes and goes on reading, kissing, smiling. He has a delicate mauve face. On his suit he has diamond and ruby buttons. In his free left hand he holds a slim ivory cane with a violet bowknot. A white lambkin peeps out of his waistcoat pocket.

Joyce's favorite chapter is "Ithaca," the "ugly duckling" of the novel. Rather repulsive when first read, "Ithaca" undergoes a curious metamorphosis with repeated readings (as does the "Cyclops" chapter, another masterpiece); a kind of hidden spirit springs to life and one begins to hear, beyond the prose surface, a truly engaging, mesmerizing voice. The narrator of "Ithaca" is ingenious: not to be confused with the moronic narrator of "Eumaeus." Despite the late hour and the fatigue of Bloom and Stephen, the narrator is fully in control of the situation, a dazzling virtuoso, almost too prodigious. Joyce said of "Ithaca" that in this chapter "All events are to be resolved into their cosmic, physical, psychical equivalents. . . . Bloom and Stephen become heavenly bodies, wanderers like the stars at which they gaze," and though these are profound and beautiful words, they refer mainly to the first part of the chapter and do not indicate the depth of the narrator's involvement with Bloom as adorer of his wife. ("He kissed the plump mellow yellow smellow melons of her rump, on each plump melonous hemisphere, in their mellow yellow furrow, with obscure prolonged provocative melonsmelonous osculation.")

"Ithaca" lent itself to continual expansion since its structure—that of a mock-catechismal question-and-answer—does not depend upon ordinary narrative movement. Movement as such is in the background, shadowy and flitting and quite unsurprising (Bloom makes Stephen hot chocolate; they sit together in Bloom's kitchen and talk; Stephen leaves); the subject of the chapter is really its technique, the narrator's skill at extracting from his heroes' particularity a sense of the general, the universal, the cosmic, that is both dwarfing and enlarging. The question-and-answer structure is by its very nature comic. In catechism books the humor is possibly unintentional: "Who made the world?" "God made the world." etc. (The young communicant memorizes questions and answers both; he comes to see that there are no questions without immediate, perfunctory answers, and of course there are no answers without proper questions.) Read as a brilliant parody of the naturalistic inclination in Joyce and others, "Ithaca" can also be read as a sympathetic commentary on Bloom's inner life, for it has been Bloom's strategy throughout the day to offer up his painful situation as representative of the human condition. In this he is, surely, a bit of an artist (as he is, also, in remembering inaccurately the effectiveness of his challenge to the anti-Semite in Barney Kiernan's): he takes on the role of Athena herself in altering his appearance for the better and thus disguising his own possibly tragic dilemma. He is, after all, a man whose father has committed suicide and whose only son has died, and he has been exiled from his immensely attractive—and "fertilisable"—wife for ten years, five months, and eighteen days. Even "mental intercourse" has failed him with Molly for the past nine months. But the narrator accommodates Bloom's stratagem by relating his situation to the general, the "objective," and in so doing underscores the absurdity of the process. The suicide note *To my Dear Son Leopold* is nearly smothered by an accumulation of miscellaneous junk—old postcards, old tickets, a recipe for the "renovation of old tan boots," a prospectus of the Wonderworker, the world's greatest remedy for rectal complaints. Bloom can also take heart that things are not worse than they are, for the narrator is quick to supply him with numerous unfortunate possibilities—

> Reduce Bloom by cross multiplication of reverses of fortune, from which these supports [i.e., certain of his financial arrangements] protected him, and by elimination of all positive values to a negligible negative irrational unreal quantity.
>
> Successively, in descending helotic order: Poverty: that of the outdoor hawker of imitation jewellery, the dun for the recovery of bad and doubtful debts, the poor rate and deputy cess collector. Mendicancy: that of the fraudulent bankrupt with negligible assets paying 1s. 4d. in the £, sandwichman, distributor of throwaways, nocturnal vagrant, insinuating sycophant, maimed sailor, blind stripling, superannuated bailiff's man, marfeast, lickplate, spoilsport, pickthank, eccentric public laughingstock seated on bench of public park under discarded perforated umbrella. Destitution: the inmate of Old Man's House (Royal Hospital), Kilmainham, the inmate of Simpson's Hospital for reduced but respectable men permanently disabled by gout or want of sight. Nadir of misery: the aged impotent disfranchised ratesupported moribund lunatic pauper.

(So brilliant is the narrator-technique of "Ithaca" that one can understand Faulkner's having appropriated it, in part, and with his own necessary alterations, as one of the "voices" of Yoknapatawpha that has come to sound most Faulknerian.)

The narrator patronizes Bloom's fascination with Woman, which is of course Joyce's as well, and one of the novel's central moments unites Bloom and Stephen as they gaze at the mystery of the "invisible" Molly, "denoted by a visible splendid sign, a lamp." It is difficult to deduce Stephen's attitude toward either Bloom or Molly. Some readers have felt that the young man is temporarily but sincerely "united" with Bloom, so that the fatherless son and the sonless father achieve a genuine intimacy, a communion; others have felt—

perhaps cynically—that Stephen declines Bloom's offer of a bed for the night because he is bored and impatient with Bloom. Certainly he has been rather curt with the older man; but Stephen is, as Buck has pointed out, an 'impossible person," not easy to love. His reaction to the photograph of Molly as a "Spanish type," proudly shown him by Bloom in the cabman's shelter, is equally difficult to interpret. He says the photograph is "handsome" but he is, perhaps, only being diplomatic; it is quite likely that he is embarrassed by her "symmetry of heaving *embonpoint*." (The subtlety of this scene calls to mind the dramatic ironies of a similar scene in Proust: lovers showing photographs of their mistresses to each other and being quite astonished at the fact that neither woman is at all attractive to the man not her lover; i.e., not deranged in his judgment by love.) In Bloom's garden, in the presence of the heaven-tree of stars hung with humid nightblue fruit, and in the presence of the invisible mystery of Woman, each is silent, "contemplating the other in both mirrors of the reciprocal flesh of theirhisnothis fellowfaces." But a moment later Stephen has gone and it seems quite likely, despite their plans, that they will not meet again. Alone, Bloom feels "the cold of interstellar space": he is once again Bloom, a quite ordinary mortal.

Moonstruck, womanstruck, Bloom has no choice but to accept his situation. His "slaying of the suitors" is his greatest stratagem—he destroys them by destroying in himself (in contrast to Odysseus' lust for violence) any desire for revenge; he makes of his wife's lovers mere heavenly bodies, impersonal, inhuman, as deluded as he himself has been, each imagining that he is the sole lover when he is, in fact, only the "last term of a preceding series." But Bloom's philosophizing is continually mocked or at least qualified by the ghostly narrator who presides over "Ithaca." Surely the list of Molly's lovers is an unlikely one—? In her soliloquy she dwells upon only a few of these men and it isn't at any time clear what her relationship has been with them—even Blazes seems to be new to her, adultery itself possibly a new experience, since she says "anyhow its done now once and for all with all the talk of the world about it people make its only the first time after that its just the ordinary do it and think no more about it." Bloom, however, teases himself with or is teased by a catalogue of lovers—

> Assuming Mulvey to be the first term of his series, Penrose, Batell d'Arcy, professor Goodwin, Julius Mastiansky, John Henry Menton, Father Bernard Corrigan, a farmer at the Royal Dublin Society's Horse Show, Maggot O'Reilly, Matthew Dillon, Valentine Blake Dillon (Lord Mayor of Dublin), Christopher Callinan, Lenehan, an Italian organgrinder, an unknown gentleman in the Gaiety Theatre, Benjamin Dollard, Simon Dedalus, Andrew (Pisser) Burke, Joseph Cuffe, Wisdom Hely, Alderman John Hooper, Dr Francis Brady, Father Sebastian of Mount Argus, a bootblack at the General Post Office, Hugh E. (Blazes) Boylan and so each and so on to no last term.

Which leaves the matter quite unclear: genuine acts are smothered by masochistic (or perhaps self-serving) fantasy.

The voice of "Ithaca" contrasts sharply with the voice of Molly in the novel's concluding chapter. Molly is matter-of-fact, not very mysterious to herself, highly perceptive in ways that relate to her marriage, frequently comic but without a sense of humor; she is an Earth Goddess of sorts, but really an ordinary heavy-set woman in her mid-thirties, whom the men of "Ithaca" have interpreted according to their own needs and who is therefore "invisible" to them. Some of Joyce's most compelling prose is launched in the service of this misapprehension; does the poetic impulse itself depend upon imaginative projection, a defiant remaking of earthy reality into something rich and strange? Joyce's well-known remark to Frank Budgen in a letter about Molly as the transpersonal *Weib* suggests that he stands with

Bloom in the garden, under the spell of the unattainable female:

> What special affinities appeared to him to exist between the moon and woman?
>
> Her antiquity in preceding and surviving successive tellurian generations: her nocturnal predominance: her satellitic dependence: her luminary reflection: her constancy under all her phases, rising, and setting by her appointed times, waxing and waning: the forced invariability of her aspect: her indeterminate response to inaffirmative interrogation: her potency over effluent and refluent waters: her power to enamour, to mortify, to invest with beauty, to render insane, to incite to and aid delinquency: the tranquil inscrutability of her visage: the terribility of of her isolated dominant implacable resplendent propinquity: her omens of tempest and of calm: the stimulation of her light, her motion and her presence: the admonition of her craters, her arid seas, her silence: her splendour, when visible: her attraction, when invisible.

Bloom and the rhetoric his infatuation inspires slip into unconsciousness; the "unusually fatiguing" day comes to a close. Consciousness itself, the activies of Dublin's men, is woven daily and then unravelled by the night, only to be taken up again the next morning . . . and unravelled again the next night. It is hardly an accident that Joyce, with his conventional mistrust of "the female," should assign to Molly the comic last word of *Ulysses*. What better way to level the pretensions of men than by having the most ordinary of Dublin voices carry us out of the novel? Molly slips into sleep and then awakes, slips into sleep and again awakes, seven times, and the eighth time falls asleep and into a dream of her seduction of Leopold Bloom ("well as well him as another"), which seems to have taken place while her mind was far distant: "I wouldnt answer first only looked over the sea and the sky I was thinking of so many things he didn't know . . ." and the novel comes to an end on a note that is both tender and comic, mildly cynical and sentimental.

❧

The impulse of comedy is gregarious; it is really comedy, rather than tragedy, that "breaks down the dykes" between people (to use Yeats's expression) and unlooses a communal music. Inherent in a multivoiced composition like *Ulysses* is the sense that individuals are most themselves when fulfilled in relationships with others: Leopold Bloom and Simon Dedalus, for instance, two "fathers" of a sort, become *Siopold* and are consumed/consummated in the most musical chapter of *Ulysses,* the "Sirens," through the power of music. It is the human voice, regardless of what it says, that is important to Joyce; he gives to Simon Dedalus a remark that was probably John Joyce's—"It [music] was the only language."

Many voices, many intonations; many narrators struggling to impose upon the world their own interpretations; an opera-like work that is best appreciated if read and re-read, aloud if possible, with an awareness of the "jocoseriousness" that underlies each passage. What does *Ulysses mean* . . . ? A fabulous artifice. A stratagem. A novel to complete the tradition of the novel. As Stephen instructs himself while trying to impress his skeptical audience in the National Library: "Local color. Work in all you know. Make them accomplices." Magnificent as it is structurally, it lives in its passages, its speeches, its moments of harmony and collision. Perhaps it is a masterpiece of gossip; Joyce might have been most pleased to hear his readers, so many decades later, still asking questions about his characters. Out of the din of so many voices there arises, irresistibly, a sense of the hilarious nature of the universe. Much is suggested, very little is actually stated. We come away from the novel as we are likely to come away from life itself, with numerous teasing, maddening, unanswerable questions—

How many lovers did Molly Bloom really have?
Who was M'Intosh?
What happened in Westland Station?—did Stephen fight with Buck Mulligan and hurt his hand?
Where does Stephen sleep after leaving Bloom?
Was Bloom really—as the mean-spirited narrator of "Cyclops" says—involved in some sort of illicit "Hungarian lottery"?
Can it be true—as Molly says—that Bloom sent Milly away so that she and Blazes could become lovers at 7 Eccles St.?
If Molly is "fertilisable" will she have another baby?
Will she really take Italian lessons from Stephen?
Whose "fault" is it that Bloom has been sexually estranged from Molly for over ten years?
Is Bloom really a Freemason?
What did Molly and Bartell d'Arcy do together on the choir stars?
Who *was* M'Intosh and who was the dead lady he loved?
Will Molly really bring Bloom eggs for breakfast?—more than once?

FINKELSTEIN MEMORIAL LIBRARY